Food Webs and the Dynamics of Marine Reefs

Food Webs and the Dynamics of Marine Reefs

Edited by
Tim R. McClanahan
George M. Branch

2008

Oxford University Press, Inc., publishes works that further
Oxford University's objective of excellence
in research, scholarship, and education.

Oxford New York
Auckland Cape Town Dar es Salaam Hong Kong Karachi
Kuala Lumpur Madrid Melbourne Mexico City Nairobi
New Delhi Shanghai Taipei Toronto

With offices in
Argentina Austria Brazil Chile Czech Republic France Greece
Guatemala Hungary Italy Japan Poland Portugal Singapore
South Korea Switzerland Thailand Turkey Ukraine Vietnam

Published by Oxford University Press, Inc.
198 Madison Avenue, New York, New York 10016

www.oup.com

Library of Congress Cataloging-in-Publication Data
Food webs and the dynamics of marine reefs / Timothy R. McClanahan
and George M. Branch, editors.
p. cm.
ISBN 978–0–19–531995–8
1. Food chains (Ecology) 2. Coral reef ecology. I. McClanahan, T. R.
II. Branch, George.
QH541.15.F66F66 2008
577.7'89—dc22 2007027165

9 8 7 6 5 4 3 2 1
Printed in the United States of America
on acid-free paper

Preface

Ecosystem management has recently become a major focus of efforts to sustain and manage natural resources. It primarily depends on an understanding of the state of the environment, its component taxa—and the interactions between them—all of which comprise the food web. Consequently, this book is an effort to facilitate the desired transition toward ecosystem management of marine reefs by summarizing some of the recent and significant advances in our understanding of the Earth's shallow subtidal marine reefs in the last few decades. It offers a current review of our understanding of shallow, benthic marine reefs and associated fisheries, focusing on food webs and how they have been-and are currently being-altered by human influences. The volume's authors have extensive experience that collectively spans the globe, and they bring together the disparate literature into a synthetic and holistic understanding of these ecosystems. The first chapter provides an overview of the development of the food-web model in marine reef ecology. The following seven chapters introduce the various environments and their food webs, describing how they differ in space and time and addressing the main human influences upon them. Based on these data and the information reviewed, the main findings of all of the authors are summarized by the editors in the final chapter, where they synthesize the commonalities and present recommendations that could potentially alleviate some of the environmental and biodiversity problems currently plaguing the Earth's shallow marine reefs.

The editors and authors would like to thank Mike Behrens, Charles Birkeland, A. Born, Monica Calvopiña, Thomas B. Clark, Paul Dayton, Luis D'Croz, Graham Edgar, Jack Engle, Eduardo Espinoza, John Field, Jared Figurski, Peggy Fong, Mike Foster, Lauren Garske, Scott Henderson, Gordon Hendler, Brian Kinlan, Steve Lonhart, Diego Lirman, Jane Lubchenco, Kathy Marten, Dayanara M. Macias Mayorga, Juan Maté, Lynn McMasters, Ann Miller, C. Moeseneder, Daniel Pauly, John Pearse, Sandie Salazar, Joseph Serafy, Scoresby Shepherd, Tyler Smith, Veronica Toral-Granda, Gunter K. Reck, Bill Robertson, and Petra Wallem for help with many aspects of the research, writing, and reviewing. Gratefully acknowledged are partnerships with and assistance from the Autoridad Nacional del Ambiente, Panama, Beneficia Foundation, CSIRO Marine and Atmospheric Research, the Conservation Science Institute, Charles

Darwin Research Center, the European Commission's INCO-DC program, Galápagos National Park Service, MEDC—the Marine Ecosystems Dynamics Consortium, Andrew W. Mellon Foundation, the Gordon and Betty Moore Foundation, the National Geographic Society, the David and Lucile Packard Foundation, Parque Nacional de Coiba, the Pew Charitable Trusts, PISCO—the Partnership for Interdisciplinary Studies of Coastal Oceans, Recursos Marinos—INRENARE (Instituto Nacional de Recursos Naturales Renovables), David H. Smith post-doctoral fellowship of The Nature Conservancy, Smithsonian Tropical Research Institute, the South African National Research Foundation, South African Network for Coastal and Oceanographic Research, University of British Columbia Fisheries Centre, the University of Cape Town, University of California Faculty Fellowship, USAID, U.S. National Science Foundation—Biological Oceanography Program, and the Wayne and Gladys Valley Foundation.

Contents

Contributors

Stuart Banks
Charles Darwin Research Station
Casilla 17-01-3891, Quito
Galápagos Islands, Ecuador
E-mail: sbanks@fcdarwin.org.ec

George M. Branch
Zoology Department
University of Cape Town
Rondebosch, Cape Town, 7701
South Africa
E-mail: gmbranch@egs.uct.ac.za

Rodrigo H. Bustamante
Northern Fisheries and Ecosystems Research Group
CSIRO Marine and Atmospheric Research
P.O. Box 120, Cleveland 4163, Queensland, Australia
E-mail: rodrigo.bustamante@csiro.au

Mark Carr
Long Marine Laboratory
100 Shaffer Road
University of California
Santa Cruz, CA 95060
E-mail: carr@biology.ucsc.edu

James A. Estes
U.S. Geological Survey
Center for Ocean Health
100 Shaffer Road
Santa Cruz, CA 95060
E-mail: jestes@cats.ucsc.edu

Jose M. Fariña
Center for Advanced Studies in Ecology and Biodiversity
Pontificia Universidad Católica de Chile
Alameda 340, Santiago, Chile
E-mail: jmfarina@bio.puc.cl

Peter W. Glynn
Rosenstiel School of Marine and Atmospheric Science
4600 Rickenbacker Cswy
University of Miami
Miami, FL 33149
E-mail: pglynn@rsmas.miami.edu

Michael Graham
Moss Landing Laboratories
8272 Moss Landing Road
Moss Landing, CA 95039
E-mail: mgraham@mlml.calstate.edu

Ben Halpern
National Center for Ecological Analysis and Synthesis
735 State Street
Santa Barbara, CA 93101
E-mail: halpern@nceas.ucsb.edu

Tim R. McClanahan
Marine Programs
Wildlife Conservation Society
Bronx, NY 10460
E-mail: tmcclanahan@wcs.org

Bruce Menge
Department of Zoology
Oregon State University
Corvallis, OR 97331-2914
E-mail: mengeb@science.oregonstate.edu

F. Patricio Ojeda
Center for Advanced Studies in Ecology and Biodiversity
Pontificia Universidad Católica de Chile
Alameda 340, Santiago, Chile
E-mail: pojeda@genes.bio.puc.cl

Thomas A. Okey
Bamfield Marine Sciences Centre
P.O. Box 100
Bamfield, BC, Canada
V0R 1B0
E-mail: tokey@bms.bc.ca

Alvaro T. Palma
Center for Advanced Studies in Ecology and Biodiversity
Pontificia Universidad Católica de Chile
Alameda 340, Santiago, Chile
E-mail: apalma@ucsc.cl

Food Webs and the Dynamics of Marine Reefs

1

Marine Food Webs: Conceptual Development of a Central Research Paradigm

Bruce Menge

Food webs, or diagrammatic descriptions of trophic connections among species in communities, have been a central focus of ecological research at least since Darwin's time (Darwin 1859). In his discussion of food webs and food chains, Elton (1927) used the term "food cycles" to refer to these diagrams, and considered questions of limits to food chain length, body size patterns, and other still-current issues. Lindeman (1942) created the field of ecological energetics, which later became ecosystem ecology, and focused the attention of ecologists on energy flows between trophic levels within webs. Both of these approaches—the "who-eats-whom" and the energy-flow perspectives—are descriptive. In the sense of indicating the presence or absence of an interaction or the pathways of energy flow through a food web, both approaches are also dynamic, but neither captures the dynamic consequences of interactions (Paine 1980, Menge and Sutherland 1987).

The pioneering experimental studies of Connell (1961a,b) and Paine (1966) and the conceptual advances of Hutchinson (1959) and Hairston and colleagues (1960) dramatically altered the focus of community ecologists. By demonstrating that species interactions could have striking affects on the community structure of rocky intertidal habitats, for example, Connell and Paine made a strong case for the power of experimentation in revealing the dynamics of natural communities. Simultaneously, these studies also demonstrated the particular advantages of marine systems as model systems for community analysis. Specifically, predation and competition were shown to strongly influence the distribution, abundance, and diversity of rocky intertidal communities in both Scotland and Washington State. Further, the experiments in each location suggested that not all species were equivalent in their ecological effect on species. Although the lessons of these results were slow to take root in terrestrial ecology, marine and freshwater community ecology was inalterably shifted from a largely descriptive research paradigm toward a perspective in which experimentation occupied a central role.

The insights of Hutchinson (1959) and Hairston and colleagues (1960) also played a pivotal role in moving community ecology into the modern era, and for both cases, food webs and food chains were central to the conceptual advances they fostered. With food webs at the core, Hutchinson (1959) knitted together a synthesis of the factors that influenced species diversity. He drew attention to the important roles of species interactions, size structure, productivity, selective food consumption by consumers, stability, niche partitioning, and habitat complexity. All of these areas became and remain foci of intense research activity.

Hairston and colleagues (1960) proposed that large-scale community pattern depends on what are now termed "trophic cascades" (Carpenter and Kitchell 1993), again launching, after a time lag beset by controversy, an entire area of intensive research. In their model, they viewed entire communities as food chains, with predators at the top trophic level determining the structure of the bottom trophic level (plants, or more generally, "basal" species) by controlling the middle trophic level (herbivores). After about two decades of sporadic debate over this idea (Murdoch 1966, Ehrlich and Birch 1967, Slobodkin et al. 1967, Fretwell 1977, Oksanen et al. 1981), empirical evidence for and against it began to accumulate (Estes et al. 1978, Strong 1992, Marquis and Whelan 1994, Polis 1994, Estes and Duggins 1995, Krebs et al. 1995, Polis and Strong 1996, Estes et al. 1998, Polis 1999, Schmitz et al. 2000, Carpenter et al. 2001, Schmitz 2003). In addition, spin-off concepts such as indirect effects (Holt 1977, Schmitt 1987, Menge 1995, Abrams et al. 1996) were also highlighted, and efforts to understand this issue continue to the present (Shurin et al. 2002, Borer et al. 2005).

Here I trace key aspects of the development of food webs and related concepts and explain their role in fostering scientific advances in community ecology. I discuss them in the context of the contribution of studies in marine ecosystems to our current understanding of the dynamics of communities, and conclude by considering future directions for research in this area.

Food-Web Theory: Natural History Gains Respect

Although food webs were always central take-off points for experimental studies of community dynamics, they were regarded by most ecologists as simple descriptive diagrams that were useful in visualizing the general trophic structure of the community of interest (Paine 1988). Most food webs were not intended, by the scientists who constructed them, to be exhaustive descriptions of all species or all linkages in a community and were typically heterogeneously resolved, often with very general groupings such as "snails" or "barnacles" or "algae," as well as highly specific ones (species), as nodes in the webs. In the late 1970s, food webs were moved dramatically from an ancillary role in field research programs to center stage in the pantheon of ecological conceptual development, when Cohen (1978) led the way to the creation of a new body of food-web theory, complete with a more rigorous set of definitions of food-web concepts (tab. 1.1). A major rationale, and hope for these efforts was that, if subsequent testing and analysis was consistent with theory, simple observation could be used to reveal community dynamics. This would obviously save a great deal of time and effort. Although by this time the power of experimental approaches for gaining insight into how communities worked was clear, it was also clear that experimentation was costly in time and research funds, and that experiments had limited capability to reveal dynamics at larger scales of space and time (Diamond 1986). Thus, if analysis of webs of trophic links among species actually revealed the dynamics underlying the structure of these webs, the pace of understanding of communities could potentially increase dramatically.

Table 1.1. Food-web concepts taken from Cohen 1978, Pimm 1982, and Schoener 1989.

Concept	Definition
Trophic species (trophospecies, node)	Set of species with same diets and same predators
Links (trophic links, edge, direct effects)	The line that connects consumer to prey
Basal species	Species at the base or bottom of the food web feeding on no other species but being fed on by others. Includes plants, and in marine systems, can include sessile invertebrates if space is regarded as the limiting resource
Intermediate species	Species that are both prey and predators
Top predator	Species feeding on basal or intermediate species with no predators of their own

(Continued)

Table 1.1. (Continued)

Concept	Definition
Trophic level	Number of links + 1 between a basal species and the species of interest.
Food chain	Path of links from a basal to a top species
Connectance	Quantitative description of the fraction of possible links that are actually present. $C = L / [S(S-1)/2]$, where L is the number of trophic links and S is the number of trophospecies.
Cycle (feeding loop)	Directed sequence of links starting and ending at the same species.
Linkage density	Average number of feeding links per species, L/S.
Omnivory	Predation on prey occurring on more than one trophic level. Can include food-chain omnivory (feeding on more than one trophic level), plant–animal omnivory (feeding on both plants and animals), live–dead omnivory (feeding on both living organisms and detritus), and life-history omnivory (feeding on different life-history stages of the same species that occur on different trophic levels)
Web vulnerability	Number of predators/prey
Web generalization	Number of prey/predator
Sink food webs	All feeding relationships of a single top predator
Source food webs	All feeding relationships arising from a single basal species
Community webs	Entire set of feeding relationships
Compartmentation	Subwebs of food webs that are not tightly linked to other subwebs. Theory suggests that compartmentation is one way diverse webs can be dynamically stable.

Cohen's (1978) work touched off a tremendous amount of theoretical and empirical analysis of food webs collected from the literature (e.g., ECOWeB) (Cohen et al. 1993). Early analyses generated a host of "general" patterns, including:

1. Food chains are consistently shorter than expected
2. Omnivory is rare
3. Compartmentation does not occur
4. Cycles are rare
5. Food chains are shorter in two-dimensional than three-dimensional habitats
6. Food webs are "interval;" that is, overlaps between species can be represented in two dimensions
7. Connections decrease with increased species richness

In addition, several patterns have emerged that are "scale invariant" (unchanging with an increase in web species richness): constancy was seen in

8. Proportions of basal, intermediate, and top species
9. Proportions of links between basal, intermediate, and top species
10. Number of links per species
11. Number of predators per prey
12. Number of prey per predator

Ideas about why such patterns emerged from these analyses included hypotheses based on:

- Low efficiency of energy transfer up food chain
- Productivity limits
- Dynamical stability (in analytical models, simpler webs are more stable)
- Habitat dimension
- "Productive space" (maximum food chain length depends on the area of space required for key components of community to persist × productivity/area; Pimm 1982, Schoener 1989)

The "scale invariance laws" were the focus of intense scrutiny, and a series of critiques cast serious doubt on their robustness (Martinez 1991, Polis 1991, Hall and Raffaelli 1993, Goldwasser and Roughgarden 1997, Bersier et al. 1999). Collectively, these studies demonstrated that when highly resolved, or when all nodes were actually species, and intensively sampled, virtually all the scale invariance disappeared. Proportions of species at different trophic levels, mean chain length, links per species (L/S), the relation between connectance and numbers of species, predator:prey ratio, prey:predator ratio, and proportions of links all varied substantially with higher resolution and more intensive sampling. The obvious conclusion was that the apparent constancy of these measures was, in fact, an artifact of low sampling intensity and poor data quality. Hence, as with the more than 20-year focus on competition theory that began in the late 1950s, these analyses dashed the early hopes that food-web analysis might prove to be a powerful shortcut to the development of understanding of community dynamics. In recent years, food-web analysis has continued, but the issue of scale invariance is largely a dead issue. New research has focused on exploring the relationships between food web and network theory (Borer et al. 2002, Dunne et al. 2002), on adding details such as abundance and body size to food-web theory (Cohen et al. 2003, Reuman and Cohen 2004) and further analysis of the dynamical consequences to model webs of connectance and diversity (Williams et al. 2002, Melian and Bascompte 2004).

Dynamical Approaches: Interaction Webs and Interaction Strength

Parallel to, and often intertwined with the activity on food-web theory and dynamics, were comparably intense efforts on experimental analysis of community dynamics, with a heavy emphasis on marine and freshwater ecosystems. The focus was on understanding what came to be termed "interaction webs," defined as the subset of species in a community linked together by strong interactions (Menge and Sutherland 1987; see also, MacArthur 1972, Paine 1980). In addition to their exclusion of trophic links that have no dynamical consequences for community structure, interaction webs also include strong nontrophic links, such as competition, facilitation, and mutualism. Interaction webs are at the heart of community models such as environmental stress models (Menge and Sutherland 1987, Menge and Olson 1990, Bertness and Callaway 1994, Menge et al. 1996, Bruno et al. 2003), which predict how biotic interactions and physical stresses should vary in importance in their effect on community structure along gradients of environmental stress.

Interaction strength, or the magnitude of the links that connect interacting species, has been a central parameter in both theoretical and empirical analyses of community dynamics. Theoretically, model communities were unstable unless they met the criterion $I\sqrt{SC} < 1$, where I is interaction strength, S is species richness, and C is connectance (May 1973). That is, if interaction strength varies, then a community of given S can have few (low C) strong interactions (high I) or many (high C) weak interactions (low I).

Although experimental studies (Connell 1961a; Paine 1966, 1974; Dayton 1971; Menge 1976; Estes et al. 1978) seemed to suggest that communities were characterized by few strong and many weak interactions, empirical efforts to quantify the distribution of interaction strength in communities did not begin until the 1990s (Paine 1992). The results of several analyses in marine and nonmarine systems suggested that earlier impressions were accurate; communities tended to have mostly weak interactions and a few strong interactions (Paine 1992, Fagan and Hurd 1994, Raffaelli and Hall 1996; but see Sala and Graham 2002). Other analyses suggested that interaction strength was context dependent; that is, a given interaction could be strong at one point along an environmental gradient or at one time and weak elsewhere or at another time (Menge et al. 1996; Navarrete and Menge 1996; Berlow 1997; 1999). These studies also touched off discussions about the extent to which empirical measures actually related to the theoretical concept of interaction strength I and which measures were most appropriate (Laska and Wootton 1998, Berlow et al. 1999, Abrams 2001).

One issue that emerged from these ideas was whether, as some advocated (Paine 1992), the community role of a species was best reflected by per capita interaction strength. This is the measure used in analytical models of communities and thus seemed to provide the best link between theoretical and empirical studies of community dynamics. Two analyses, however, suggested that reliance on just per capita interaction strength to judge the community role of a species can be misleading. For example, in their studies of interaction strengths of predators on prey in a mudflat community in Scotland, Raffaelli, and Hall (1996) showed that shorebirds had very high per capita interaction strengths. Use of different indices of interaction strength did not alter this pattern (Berlow et al. 1999). However, the strengths of these effects were greatly affected by predator density, which was very low (Raffaelli and Hall 1996). Since calculation of empirically-based measures of interaction strength use the density of predators in the denominator, and since shorebird densities per unit area were very low ($0.004/m^2$), the high per capita rate was a result of high sensitivity to division by very small numbers. Hence, even though shorebirds can feed at a very high rate, their density relative to prey density was low and their population effect as predators was not strong.

In another example involving two invertebrate predator species, Navarrete and Menge (1996) found that per capita interaction strengths of sea stars feeding on mussels varied among sites but their population effects were relatively consistent among sites, despite wide among-site variation in sea star density. In contrast, whelk per capita interaction strengths were less variable among sites, and their population effect varied mostly as a function of their density. Navarrete and Menge (1996) suggested that these differences were indicative of between-species differences in functional and numerical responses. Sea stars have higher per capita interaction strength than do whelks, and are thus more voracious predators, but as predator density increases, their population effect saturates and per capita rates decline as their density increases. Whelks, in contrast, feed more slowly and thus have a weak functional response to prey, so population effects of whelks will vary linearly with predator density rather than saturate, as whelks respond to prey primarily through a numerical response.

The analyses summarized above emphasize the importance of strong interactors in the study of community dynamics. Are "weak" interactors unimportant to community dynamics? Is a focus on only strongly interacting species appropriate when evaluating how communities work, or when making recommendations regarding ecosystem management? Berlow (1999) examined this question in a whelk–barnacle–mussel interaction web. Using experiments repeated in successive years with different intensities of prey recruitment, he demonstrated that when prey recruitment is low, a weakly interacting predator could have a strong effect. Further, even when predation is on average weak, interaction strength can be highly variable: in some locations in the environmental mosaic, predation will be strong and in other locations predation will be weak. Thus, in the context of variable environments (a "normal" condition in most ecological communities), weakly interacting species still may be a critical component of communities. Conditions such as sharp reductions in abundance of strongly interacting species may foster conditions under which previously weakly interacting species become an important and strongly influential component of the system. In other words, a wide range of species-interaction strengths may enhance the resilience of communities to perturbations.

Critical Species

In addition to demonstrating the importance and range of interaction strength, experimental analyses of community dynamics have revealed the existence in communities of particular types of species that play key roles in community dynamics. Keystone species are probably the most prominent example (Paine 1966, Power et al. 1996, Menge and Freidenburg 2001), but other kinds of species have also been labeled as critically important to the dynamics of communities. For example "dominants" have been defined as species that have a strong influence on communities because of their high abundance (Power et al. 1996). "Key-industry" species have been identified as prey of intermediate trophic status that support a large group of consumers (Elton 1927, Birkeland 1974). "Foundation" species are the "group of critical species that define much of the structure of a community" (Dayton 1972). "Ecosystem engineers" are species that modify, maintain, or create habitats, thereby controlling the availability of resources to other organisms (Jones et al. 1997). Clearly, these different ideas overlap in various ways, but all share the notion that not all species are functionally or dynamically equivalent.

The idea behind the keystone-species concept has been present in ecological thought probably for over a century (Paine 1995). The term was originally coined to identify species high in the food web that can greatly modify the composition and physical appearance of an ecological community through its interaction with prey (Paine 1969). The concept eventually enjoyed wide application, both in basic and applied ecology (Mills et al. 1993), but also attracted criticism (Mills et al. 1993, Hurlbert 1997). The primary concern was that the term had become so widely applied to any strongly interacting species that the concept was of questionable usefulness. There were also concerns about how to quantify the concept, a step that would greatly increase its utility, and about the growing realization that the community effects of keystone species were context-dependent. That is, while a given consumer might have a clear "keystone" role in some parts of its range, in other parts its role might be relatively minor (Paine 1980, Menge et al. 1994).

As a consequence of these issues, efforts to refine or restrict the definition of keystone species were made (Power et al. 1996, Menge and Freidenburg 2001). Consensus has been elusive, however, and as with many ecological ideas, definitions must usually be provided with each usage to make clear the context in which the term is used. Here I follow terminology recommended in a previous review (Menge and Freidenburg 2001): *a keystone species is a consumer having a disproportionately large effect on communities and ecosystems.* By this definition I include all consumers and exclude plants, sessile animals, or "resources," such as carcasses, habitat services, and nutrients. I further note that keystone species are just one type of strong interactor, and that although a community may have a single keystone species, it may have many strong interactors or "critical" species.

Are keystone species predictable? Can they be identified without experimentation? At this point, it seems clear that the answer to both questions is no. As discussed above, it is possible to observe feeding activities, and from this, to generate hypotheses about the likely role of a consumer, but in my judgment, it is currently impossible to identify a keystone predator a priori. Nor does it seem possible to predict whether a particular community or ecosystem will be characterized by keystone predation, "diffuse" predation (such as by a group of predatory species rather than by a single dominant one), or weak predation.

Two examples of how difficult it can be to identify a species' role include:

- In South Africa, simple observation in the subtidal communities of Marcus and Malgas Islands provided no clue to the incredibly strong effect of the whelk *Burnupena papyracea* on lobsters *Jasus lalandii* (see chapter 3). Only by transplanting lobsters from Malgas, where they were abundant, to Marcus, where they were absent, did the investigators learn that at high density *Burnupena* can turn the tables and become a predator of its predator (Barkai and McQuaid 1988).
- In Alaska, prior research had suggested that sea otters *Enhydra lutra* were keystone predators in kelp communities (Estes et al. 1978). In the 1990s, sea otters suddenly began disappearing, and only a few lucky sightings of predation on sea otters by killer whales (*Orca*) allowed the insight that enabled researchers to propose the hypothesis that changes in ecosystem dynamics had forced orcas to redirect their feeding activities on sea otters, triggering massive changes in kelp communities (Estes et al. 1998; chapter 2).

Many other examples could be added to these two. My point is that—as many others before me have noted—when manipulated, food webs, communities, and ecosystems commonly reveal dramatic and surprising dynamics, and the capacity for predicting such changes remains limited. Nonetheless, the existence of critical species is real; most communities seem to have them, and the dynamics of communities can only be understood if we understand their effects.

Prediction of whether or not a community will have a keystone species, or an ecological engineer, and identifying those species in the system that might fill this role seems elusive at this point. Nonetheless, it still seems reasonable to search for those system characteristics that might allow such insights. Efforts to do so, however, soon run into another major roadblock. Rarely do we have the knowledge we need at the scale of the ecosystem to categorize ecological traits in different communities. Even less frequently do we understand the types or magnitudes of linkages between communities and ecosystems or between adjacent ecosystems that would allow insights into the role of physical conditions, productivity, or larval supply in generating conditions that might lead to the evolution and persistence of a keystone or other type of critical species.

The Interface Between Communities and Ecosystems

Ecological communities are arrayed along environmental gradients (Whittaker 1970). In coastal marine ecosystems, key gradients include those in environmental stress, nutrients, and productivity and propagule supply (Connell 1975, Menge and Sutherland 1987, Roughgarden et al. 1988, Menge and Olson 1990, Menge et al. 1996). The effects of factors varying along these gradients can span a wide range of spatial scales, but their greatest effects seem concentrated across a smaller range of scales. Environmental stresses, such as wave forces or thermal stress, cause complex environmental gradients whose effects can span a large range of spatial scales (fig. 1.1), but these types of stresses also have effects that are apparent and readily studied on more local scales (Dayton 1971, 1975, Menge 1976, Lubchenco and Menge 1978). In contrast, bottom-up effects and propagule abundance, which also vary at a wide range of scales, may exert their strongest effects on the scales over which major coastal current ecosystems vary (fig. 1.1). Below we examine these ideas.

Environmental Stress Gradients

Like plant ecologists, marine ecologists were long impressed with the apparent influence of waves, thermal stress, and other effects

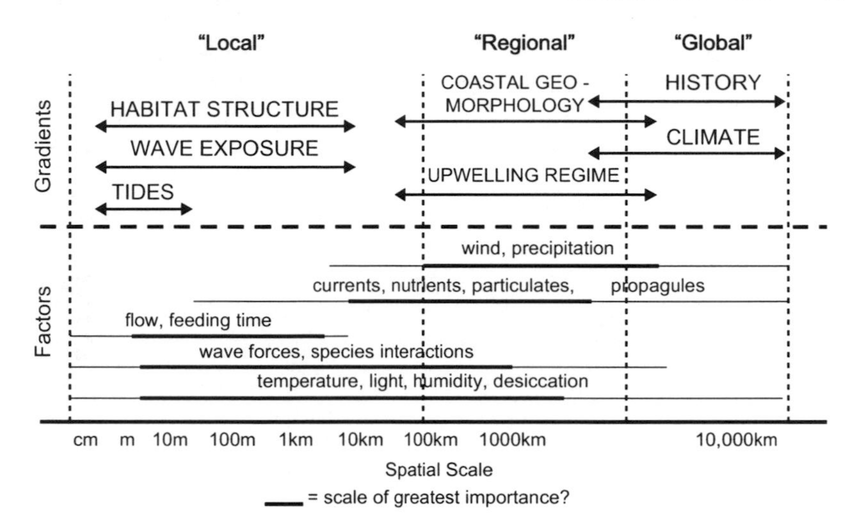

FIGURE 1.1 Conceptual diagrams of how environmental factors vary along gradients and spatial scale. Ranges over which the different gradients and factors operate are suggested by the thin black lines, and the scales over which the different factors are judged to be most important in terms of their influence on community and ecosystem dynamics are shown by the thick black lines.

that varied with the proximity of a shore to the direct effects of ocean swell and the duration of immersion (Lewis 1964, Stephenson and Stephenson 1972). While earlier workers such as Lewis (1964) emphasized the importance of physical versus biological processes in determining patterns of zonation, patchiness, and abundance, the work of Connell (1961a,b), Paine (1966, 1974), and many others convincingly demonstrated that biological processes were responsible for at least some of those patterns initially thought to be due to physical processes. Thus, it became clear that more complete understanding would depend on integrating biological and physical processes into a more general conceptual framework.

Environmental stress models Experimental analyses of the effects of species interactions and how these varied along gradients of environmental stress (Dayton 1971, Menge 1976) spurred the development of a conceptual framework that integrated the ecological effects of physical and physiological disturbance, competition, and predation on community structure, and how these factors vary along environmental stress gradients (Connell 1975, Menge and Sutherland 1976, 1987). In the intertidal, such gradients can be complex. For example, wave stress and thermal stress gradients can run in opposite directions, such that wave force stresses can inhibit predators and remove prey via physical disturbance at the wave-exposed part of the gradient while thermal stress can also inhibit predators and kill prey via physiological disturbance at the wave-protected end of the gradient (Menge and Sutherland 1987). Such complexity can be daunting, and many have avoided or resisted efforts to develop general conceptual frameworks for how these factors interact to produce community structure. Others, however, have embraced such complexity, reasoning that complex systems are not likely to be well understood either by simple single-factor models or by asserting that the uniqueness of each system necessarily defeats any attempt at achieving generality (chapter 9).

The first environmental stress models examined the roles of physical disturbance, competition, and predation along a gradient of environmental stress (Connell 1975, Menge and Sutherland 1976, 1987). These models suggested that with increasing stress, the dominant forces controlling community structure were consumption (predation and competition), then competition alone, and finally direct effects of physical or physiological stress. The proposed mechanism was differential tolerance of mobile and sessile organisms to stress: sessile organisms were predicted to be more tolerant to stress than mobile organisms because their sessile life style denied them the option of moving away from stressful conditions. Thus, it was reasoned that sessile organisms would be under strong selective pressure to evolve higher stress tolerance.

Evidence consistent with these models and some of their assumptions came from field studies in marine benthic environments (Menge and Farrell 1989, McClanahan 1992), bog lakes (Arnott and Vanni 1993), and studies of the biomechanical properties of intertidal organisms (Denny 1988). However, studies from terrestrial habitats raised the possibility that some sessile organisms might be more, not less, susceptible to stress than mobile consumers (Louda and Collinge 1992). One situation where this might be true is when consumers are much smaller than their prey organisms, such as aphids feeding on plants. In marine environments, sessile organisms include both seaweeds and invertebrates such as mussels, barnacles, anemones, tunicates, corals, and sponges. Consumers of these can include organisms that range from being large relative to prey, such as sea stars versus barnacles and some limpets versus algae, to small relative to prey, such as some other limpets versus algae and isopods versus algae.

Revisions to environmental stress models These considerations led to a revised environmental stress model (ESM) in which the two alternatives of susceptibility to stress generated two alternative models: the consumer stress model (CSM—consumers are more susceptible than prey = the original ESM) and the prey stress model (PSM—consumers are less susceptible than prey) (Menge and Olson 1990). Evidence consistent with the PSM was obtained in experimental studies of a limpet–macroalga interaction (Olson 1992). Grazing by the limpet *Lottia digitalis* on the red turf-forming alga *Mazzaella parksii* was higher under conditions that stressed the alga than under experimentally alleviated levels of stress. In this case, the limpet was smaller than the macroalga, and sought shelter in the shade beneath the alga.

An explicit test of the CSM model, including an effort to change the limpet–alga interaction from a PSM to a CSM scenario was only partially consistent with model predictions (Menge et al. 2002a). In a whelk–barnacle system (consumer larger than prey, predicted to be an example of the CSM), feeding rates were reduced by thermal stress that was manipulated by artificial shades, but were elevated by warmer climatic conditions. In a limpet–alga system constrained to fit the circumstances where the CSM might apply (i.e., limpets grazing pieces of macroalgae rather than the entire thallus of the macroalga), feeding rates were also reduced by thermal stress, but as with the whelks, were elevated by warmer climatic conditions. Clearly the responses of consumers to stress can vary with the spatial and temporal scale of the stress. Further investigation of these relationships would be welcome.

The outcome of experimental studies in plant-dominated alpine and salt-marsh habitats led to further modification of environmental stress models (Bertness and Callaway 1994). For example, in salt marshes, experiments showed that under more stressful conditions, facilitation by plants more tolerant to salinity stress could encourage persistence of less tolerant plants (Bertness and Ellison 1987, Bertness and Hacker 1994, Hacker and Bertness 1995). This approach was extended to New England rocky intertidal habitats, showing that the frequency of facilitation was higher in more stressful high intertidal habitats than in low intertidal habitats (Bertness et al. 1999). More recently, Burnaford (2004) demonstrated that seaweeds, even in the low intertidal zone, could alleviate the effects of thermal stress on both herbivores and understory algae.

Facilitation can also occur under conditions of low environmental stress. In subtidal reefs, Hay (1986) demonstrated that becoming epiphytes on larger algae could enhance growth and survival of algae of small stature. Hay observed that small algae such as *Hypnea* sp. could persist in a system where grazing by herbivorous fish would normally remove them by settling into the axils of branches of the larger *Sargassum*. These locations provided a refuge in space from the fish, allowing persistence of the smaller alga and thereby maintaining higher community diversity. Hay (1986) termed this indirect effect of grazers on algae an "associational defense."

Synthesizing these and similar results demonstrating associational defenses with the stress-engendered facilitation documented by Bertness and Hacker (1994), Bruno and colleagues (2003) proposed a revised ESM building on the model of Menge and Sutherland (1987). Bruno and colleagues (2003) suggested that the influence of facilitation in structuring communities be added to the effects of disturbance, competition, and consumer pressure. The effect of facilitation was predicted to be highest at the high and low ends of the environmental stress gradient and lowest in the intermediate level of stress, where competition prevails. A major issue awaiting resolution is the magnitude of these postulated effects of facilitation relative to the effects of disturbance and consumption. Facilitation could range in importance from high, being a major determinant of community structure under high or low stress, or low, being a minor determinant of community structure. Testing these ideas is an area ripe for future research.

Stress and the functional role of diversity Appreciation of the importance of environmental stress gradients in structuring communities and ecosystems has expanded greatly with a recent focus on the importance of diversity in providing resistance to stress (Tilman and Downing 1994). Results from experimental grassland ecosystems have suggested that more diverse systems are more resistant to drought stress than are less diverse systems. One area of controversy revolves around whether such an effect results directly from the diversity of plants in the system or from the characteristics of key species, more of which are likely to be included in a diverse ecosystem (Allison 1999).

Attempts to test these ideas in marine communities have been few, but a recent study indicates that stress resistance is likely to be more dependent on characteristic species than on species or taxon richness or diversity. In an algal-dominated community, Allison (2004) showed that the response of the community to stress depended on the magnitude of stress, and importantly, more on species composition than on species diversity. The importance of species composition in determining biodiversity–ecosystem function relationships has also been emphasized in other studies as well, including studies that manipulated both plants and consumers (Hooper and Vitousek 1997, Duffy et al. 2001, Downing and Leibold 2002, Duffy 2002).

Productivity Gradients

The supply of nutrients varies in space and time, and can underlie gradients of productivity. This issue has long been a dominant theme in terrestrial and freshwater communities and in fisheries, but its influence in structuring communities and ecosystems in coastal marine environments has been less studied (Menge 1992). Conceptual frameworks for such effects have emphasized contrasting dynamics.

Tilman focused on the importance of nutrient gradients in structuring plant communities, thereby emphasizing the dominance of "bottom-up" effects as determinants of community structure (Tilman 1982). Building on the seminal ideas of Hairston and colleagues (1960), Fretwell (1987) and Oksanen and colleagues (1981), he envisioned productivity gradients as factors underlying the assembly of communities and thereby generating variation in consumer pressure. Their model thus proposed that community structure and dynamics reflected an interaction between bottom-up and top-down effects, with the importance of these effects alternating with increasing productivity. This alternating effect depends on herbivory and carnivory being the primary forces of consumption; if omnivory is strong, consumption should steadily increase with increasing productivity (Menge et al. 1996).

Linking benthic and pelagic marine ecosystems With respect to extrinsic factors that can drive their structure, benthic marine ecosystems can be linked to their watery milieu by three factors: inputs of nutrients, propagules, and particulates, including phytoplankton and detritus. Such inputs have been termed "ecological subsidies" (Polis and Hurd 1996). Of these factors, propagule dispersal and delivery can potentially range over enormous distances and the rates of these can be important determinants of community structure (Gaines and Roughgarden 1985, Menge and Sutherland 1987, Roughgarden et al. 1988, Connolly and Roughgarden 1999). Since the ocean currents that transport propagules are predictable on at least some scales, assessing the influence of propagule input is potentially tractable in a research program, and considerable progress has been made in examining the role of recruitment in structuring marine communities (Morgan 2001, Schiel 2004). Examining the community and ecosystem consequences of variation in nutrients and particulates, however, has lagged, principally because of logistical and scale issues. Recent efforts, reviewed in some detail elsewhere (Menge 2000, 2003b, Schiel 2004) suggest that these two factors can also have important effects on food-web structure.

Studies from several locations around the globe suggest that external inputs and their influence on marine benthic food webs are commonly driven by coastal upwelling regimes. In South Africa, Chile, the Galápagos Islands, New Zealand, and the west coast of North America, upwelling influences on the delivery of nutrients, propagules, and particulates have been linked to large-scale variation in community structure (Bustamante et al. 1995a,b; Bustamante and Branch 1996; Menge et al. 1997, 2002b, 2003, 2004; Broitman et al. 2001; Connolly et al. 2001; Nielsen 2001; Witman and Smith 2003; Worm et al. 2002; Menge 2003a; Nielsen 2003; Nielsen and Navarrete 2004; Schiel 2004; Navarrete et al. 2005). Collectively, these studies suggest that spatial variation in food-web structure and dynamics is commonly associated with variation in the coastal upwelling regime on the scale of one to thousands of kilometers. For example, in Chile variation in the dominance of mussels and barnacles is associated with a major shift at about 32°S in the large-scale upwelling pattern from weaker upwelling in the south to stronger upwelling in the north (Navarrete et al. 2005). Shifts in recruitment rates and phytoplankton concentration along the North American west coast and around the coast of the South Island of New Zealand are also associated with major discontinuities in upwelling regime (Connolly et al. 2001; Menge et al. 1997, 2003, 2004). Evidence suggests that weaker or more intermittent upwelling can generate conditions favoring high rates of delivery of propagules and particulates, and that these differences in rates of input of subsidies may underlie differences in community structure, including abundances of some dominant space occupiers, and food-web dynamics, including growth rates of key dominants, rates of recovery from disturbance, rates of predation, and competition (Guichard et al. 2003, Menge 2003a, Menge et al. 2004, Navarrete et al. 2005).

Temporal variation in upwelling can also influence food-web dynamics. In the Galápagos, Vinueza and colleagues (2005) showed that the arrival of El Niño heralded a discontinuation of upwelling, with an increase in sea temperature, but a decline in nutrients. This rippled through the food web, with primary productivity dropping and palatable algae being replaced by unpalatable species, and herbivorous crabs and iguanas declining or suffering loss of condition.

Discussion

Food-web research in benthic marine habitats has contributed substantially to the conceptual development of ecology throughout its history, and this contribution continues unabated. Models have grown from largely population-focused in the early stages of conceptual development, to foci on effects of species interactions and disturbance, on how these processes interact with physical environmental gradients and productivity gradients at the local scale, and most recently on such gradients at the meso- and macro-scales. With these developments has grown an increasing appreciation for the importance of scales in space and time, and how the influence of a process can vary at different scales. The most recent efforts have begun to address the issue of the extent to which marine communities are linked, both to one another and to the pelagic ecosystem that bathes them.

As in all of ecology, advances and increases in the complexity of conceptual frameworks present new challenges. For example, evaluating the relative importance of different processes in determining food web, community, and ecosystem structure has always been difficult. Even greater challenges come with efforts to incorporate the influences of oceanographic-scale factors such as nutrients, propagules, and particulates in determining local to mesoscale community patterns. These factors can vary at several spatial and temporal scales, and not just at the largest scales. Assessing such variation in an ecologically meaningful framework often requires new approaches and new techniques. Community variation also arises from local-scale events such as species interactions, localized propagule delivery, and storm-generated disturbance modified by headlands, offshore reefs or sandbars, shore steepness or orientation, sedimentation, and other factors. Determining the relative influence of processes varying on these different scales is a daunting task that will require, or be greatly improved, by the development of new analytic and computational approaches.

New technology has had a major effect on our ability to assess this complexity. For instance, field deployable, durable, and relatively inexpensive remote sensing devices have dramatically altered our ability to quantify physical and some biological environmental characteristics at temporal and spatial scales that are more closely matched to the ecological scales required for our research. Such advances lie at the heart of efforts to establish networks of environmental sensors that can provide a physical environmental context for future studies. These advances will provide opportunities to test predictions of conceptual ecological models in more rigorous and more direct ways than has been possible.

Linking ecological pattern and process to these better-quantified physical environmental contexts is in some ways the greatest remaining challenge to the study of community and ecosystem dynamics. One approach that is increasingly employed in this effort is the "comparative-experimental" approach, where identically designed, usually small-scale experiments are carried out across spatial and temporal gradients. For example, herbivore exclusion–nutrient addition experiments can be done in different zones across multiple sites spanning upwelling and wave exposure gradients from local to mesoscale to macroscale spatial scales. Repetition of these sorts of experiments at different temporal scales could provide insight into how the magnitude of the tested effects varies in time. Another wrinkle to such experiments would be to systematically vary the plot size in combination with all the other treatments to obtain insight into the influence of the size of the area examined within each experimental unit.

A final, somewhat intangible influence on progress in understanding the factors driving food-web structure and dynamics is the level of engagement by the ecological community in efforts to seek general principles. Although most scientists will enthusiastically embrace the position that advancement comes fastest when research is hypothesis- or question-driven, and takes place in the context of some sort of conceptual framework, in reality not all buy fully into this approach. Instead, what might be called a strict local-scale approach is adopted. The investigator is most impressed with differences among sites that are often close together, and observes that these differences can be greater than differences between sites that are farther apart. Or the results of repeated experiments might vary, which suggests that temporal variation on the local scale is of overwhelming importance. The inference is made that local spatial and temporal variation is so great that other, larger- or longer-scale influences are not important, not detectable or not approachable. One consequence might be the conclusion that understanding is possible only at the local scale, leading to the inference that broader understanding is possible only if we study every site, or alternatively that broader understanding is not possible. This is the "my site is different from your site" philosophy, implying that it is fruitless to attempt to understand similarities in pattern–process linkages in even geographically different representatives of the same type of habitat. In such cases, a final inference from this line of thought is that ecology is not a true science, or that it is a science where general principles cannot be detected through the scientific method, which is a view that regrettably accords with the views of some of our nonecological colleagues.

I most definitely do not embrace this latter vision. Instead, I believe that the search for general organizing principles in ecology has been fruitful and will continue to be so, and that only by adopting this approach do we have a hope of understanding how ecosystems will respond to disruptions on multiple scales. We must grapple with complexity to understand it, and the forces that generate it. If we do not, we risk the scrapheap of irrelevancy, and increasing marginalization in the scientific community. The chapters that follow in this book are examples of efforts to seek out and synthesize these general principles by comparing various ecosystems from disparate areas of the world.

I conclude on a positive note; we are presently at an exciting time in marine community ecology. A new generation of question-oriented ecologists is emerging with a broader and more sophisticated toolbox. They are computationally, analytically, and technologically literate, trained in increasingly interdisciplinary programs, and more willing than previous generations of ecologists to work collaboratively rather than competitively. We are already seeing early consequences of the emergence of this new generation in the form of studies that tackle problems of scales and magnitudes that were largely unimaginable a few years ago. I believe that future progress in marine community and ecosystem ecology is critically dependent on this new way of doing science, and look forward to the next decade of progress in understanding how ecosystems work.

References

Abrams, P., B.A. Menge, G.G. Mittelbach, D. Spiller, and P. Yodzis. 1996. The role of indirect effects in food webs. Pages 371–395 *in* G.A. Polis and K.O. Winemiller, editors, Food webs: integration of pattern and dynamics. Chapman and Hall, N.Y.

Abrams, P.A. 2001. Describing and quantifying interspecific interactions: a commentary on recent approaches. Oikos **94**:209–218.

Allison, G. W. 1999. The implications of experimental design for biodiversity manipulations. American Naturalist **153**:26–45.

Allison, G. 2004. The influence of species diversity and stress intensity on community resistance and resilience. Ecological Monographs **74**:117–134.

Arnott, S.E., and M.J. Vanni. 1993. Zooplankton assemblages in fishless bog lakes: influence of biotic and abiotic factors. Ecology **74**:2361–2380.

Barkai, A., and C.D. McQuaid. 1988. Predator-prey role reversal in a marine benthic ecosystem. Science **242**:62–64.

Berlow, E.L. 1997. From canalization to contingency: historical effects in a successional rocky intertidal community. Ecological Monographs **67**:435–460.

Berlow, E.L. 1999. Strong effects of weak interactors in ecological communities. Nature **398**:330–334.

Berlow, E.L., S.A. Navarrete, M.E. Power, B.A. Menge, and C. Briggs. 1999. Quantifying variation in the strengths of species interactions. Ecology **80**:2206–2224.

Bersier, L.-F., P. Dixon, and G. Sugihara. 1999. Scale-invariant or scale-dependent behavior of the link density property in food webs: a matter of sampling effort? American Naturalist **153**:676–682.

Bertness, M.D., and R. Callaway. 1994. Positive interactions in communities. Trends in Ecology and Evolution **9**:191–193.

Bertness, M.D., and A.M. Ellison. 1987. Determinants of pattern in a New England salt marsh plant community. Ecological Monographs **57**:129–147.

Bertness, M.D., and S.D. Hacker. 1994. Physical stress and positive associations among marsh plants. American Naturalist **144**:363–372.

Bertness, M.D., G.H. Leonard, J.M. Levine, P.R. Schmidt, and A.O. Ingraham. 1999. Testing the relative contribution of positive and negative interactions in rocky intertidal communities. Ecology **80**:2711–2726.

Birkeland, C. 1974. Interactions between a sea pen and seven of its predators. Ecological Monographs **44**:211–232.

Borer, E.T., K. Anderson, C.A. Blanchette, B.R. Broitman, S.D. Cooper, B.S. Halpern, E.W. Seabloom, and J.B. Shurin. 2002. Topological approaches to food web analyses: a few modifications may improve our insights. Oikos **99**:397–401.

Borer, E.T., E.W. Seabloom, J.B. Shurin, K.E. Anderson, C.A. Blanchette, B.R. Broitman, S.D. Cooper, and B.S. Halpern. 2005. What determines the strength of a trophic cascade? Ecology **86**:528–537.

Broitman, B.R., S.A. Navarrete, F. Smith, and S.D. Gaines. 2001. Geographic variation of southeastern Pacific intertidal communities. Marine Ecology Progress Series **224**:21–34.

Bruno, J.F., J.J. Stachowicz, and M.D. Bertness. 2003. Inclusion of facilitation into ecological theory. Trends in Ecology and Evolution **18**:119–125.

Burnaford, J.L. 2004. Habitat modification benefits a consumer: effects of an algal canopy on a low intertidal herbivore. Ecology **85**:2837–2849.

Bustamante, R.H., and G.M. Branch. 1996. The dependence of intertidal consumers on kelp-derived organic matter on the west coast of South Africa. Journal of Experimental Marine Biology and Ecology **196**:1–28.

Bustamante, R.H., G.M. Branch, and S. Eekhout. 1995a. Maintenance of an exceptional intertidal grazer biomass in South Africa: subsidy by subtidal kelps. Ecology **76**:2314–2329.

Bustamante, R.H., G.M. Branch, S. Eekhout, B. Robertson, P. Zoutendyk, M. Schleyer, A. Dye, N. Hanekom, D. Keats, M. Jurd, and C. McQuaid. 1995b. Gradients of intertidal primary productivity around the coast of South Africa and their relationships with consumer biomass. Oecologia **102**:189–201.

Carpenter, S.R., J.J. Cole, J.R. Hodgson, J.F. Kitchell, M.L. Pace, D. Bade, K.L. Cottingham, T.E. Essington, J.N. Houser, and D.E. Schindler. 2001. Trophic cascades, nutrients, and lake productivity: whole-lake experiments. Ecological Monographs **71**:163–186.

Carpenter, S.R., and J.F. Kitchell. 1993. The trophic cascade in lakes. Cambridge University Press, Cambridge, U.K.

Cohen, J.E. 1978. Food webs and niche space. Princeton University Press, Princeton, NJ.

Cohen, J.E., R.A. Beaver, S.H. Cousins, D.L. DeAngelis, L. Goldwasser, K.L. Heong, R.D. Holt, A.J. Kohn, J.H. Lawton, N.D. Martinez, R.E. O'Malley, L.M. Page, B.C. Patten, S.L. Pimm, G.A. Polis, M. Rejmanek, T.W. Schoener, K. Schoenly, W.G. Sprules, J.M. Teal, R.E. Ulanowicz, P.H. Warren, H.M. Wilbur, and P. Yodzis. 1993. Improving food webs. Ecology **74**:252–258.

Cohen, J.E., T. Jonsson, and S.R. Carpenter. 2003. Ecological community description using the food web, species abundance, and body size. Proceedings of the National Academy of Science USA **100**:1781–1786.

Connell, J.H. 1961a. Effects of competition, predation by *Thais lapillus*, and other factors on natural populations of the barnacle *Balanus balanoides*. Ecological Monographs **31**:61–104.

Connell, J.H. 1961b. The influence of interspecific competition and other factors on the distribution of the barnacle *Chthamalus stellatus*. Ecology **42**:710–723.

Connell, J.H. 1975. Some mechanisms producing structure in natural communities: a model and evidence from field experiments. Pages

460–490 *in* M.L. Cody and J.M. Diamond, editors, Ecology and evolution of communities. Belknap Press, Cambridge, Mass.

Connolly, S.R., B.A. Menge, and J. Roughgarden. 2001. A latitudinal gradient in recruitment of intertidal invertebrates in the northeast Pacific Ocean. Ecology **82**:1799–1813.

Connolly, S.R., and J. Roughgarden. 1999. Theory of marine communities: competition, predation, and recruitment-dependent interaction strength. Ecological Monographs **69**:277–296.

Darwin, C.R. 1859. The origin of species by means of natural selection. First edition. John Murray, London.

Dayton, P.K. 1971. Competition, disturbance, and community organization: the provision and subsequent utilization of space in a rocky intertidal community. Ecological Monographs **41**:351–389.

Dayton, P.K. 1972. Toward an understanding of community resilience and the potential effects of enrichments to the benthos at McMurdo Sound, Antarctica. Pages 81–96 *in* B.C. Parker, editor, Proceedings of the colloquium on conservation problems in Antarctica. Allen Press, Lawrence, Kans.

Denny, M.W. 1988. Biology and the mechanics of the wave-swept environment. Princeton University Press, Princeton, N.J.

Diamond, J. 1986. Overview: laboratory experiments, field experiments, and natural experiments. Pages 3–22 *in* J. Diamond and T.J. Case, editors, Community ecology. Harper and Row, N.Y.

Downing, A.L., and M.A. Leibold. 2002. Ecosystem consequences of species richness and composition in pond food webs. Nature **416**:837–841.

Duffy, J.E. 2002. Biodiversity and ecosystem function: the consumer connection. Oikos **99**:201–219.

Duffy, J.E., K.S. MacDonald, J.M. Rhode, and J.D. Parker. 2001. Grazer diversity, functional redundancy, and productivity in seagrass beds: an experimental test. Ecology **82**:2417–2434.

Dunne, J.A., R.J. Williams, and N.D. Martinez. 2002. Food-web structure and network theory: the role of connectance and size. Proceedings of the National Academy of Science USA **99**:12917–12922.

Ehrlich, P., and L.C. Birch. 1967. The "balance of nature" and "population control." American Naturalist **101**:97–107.

Elton, C.S. 1927. Animal ecology. Sidgwick and Jackson, Ltd., London.

Estes, J.A., and D.O. Duggins. 1995. Sea otters and kelp forests in Alaska: generality and variation in a community ecological paradigm. Ecological Monographs **65**:75–100.

Estes, J.A., N.S. Smith, and J.F. Palmisano. 1978. Sea otter predation and community organization in the western Aleutian Islands, Alaska. Ecology **59**:822–823.

Estes, J.A., M.T. Tinker, T.M. Williams, and D.F. Doak. 1998. Killer whale predation on sea otters linking oceanic and nearshore ecosystems. Science **282**:473–476.

Fagan, W.F., and L.E. Hurd. 1994. Hatch density variation of a generalist arthropod predator: population consequences and community impact. Ecology **75**:2022–2032.

Fretwell, S.D. 1977. The regulation of plant communities by food chains exploiting them. Perspectives in Biology and Medicine **20**:169–185.

Fretwell, S.D. 1987. Food chain dynamics: the central theory of ecology? Oikos **50**:291–301.

Gaines, S.D., and J. Roughgarden. 1985. Larval settlement rate: a leading determinant of structure in an ecological community of the marine intertidal zone. Proceedings of the National Academy of Sciences USA **82**:3707–3711.

Goldwasser, L., and J. Roughgarden. 1997. Sampling effects and the estimation of food-web properties. Ecology **78**:41–54.

Guichard, F., P. Halpin, G.W. Allison, J. Lubchenco, and B.A. Menge. 2003. Mussel disturbance dynamics: signatures of oceanographic forcing from local interactions. American Naturalist **161**:889–904.

Hacker, S.D., and M.D. Bertness. 1995. Morphological and physiological consequences of a positive plant interaction. Ecology **76**:2165–2175.

Hairston, N.G., F.E. Smith, and L.B. Slobodkin. 1960. Community structure, population control, and competition. American Naturalist **94**:421–425.

Hall, S.J., and D.G. Raffaelli. 1993. Food webs: theory and reality. Advances in Ecological Research **24**:187–239.

Hay, M.E. 1986. Associational plant defenses and the maintenance of species diversity: turning competitors into accomplices. American Naturalist **128**:617–641.

Holt, R.D. 1977. Predation, apparent competition, and the structure of prey communities. Theoretical Population Biology **12**:197–229.

Hooper, D.U., and P.M. Vitousek. 1997. The effects of plant composition and diversity on ecosystem processes. Science **277**:1302–1305.

Hurlbert, S.H. 1997. Functional importance vs. keystoneness: reformulating some questions about theoretical biocenology. Australian Journal of Ecology **22**:369–382.

Hutchinson, G.E. 1959. Homage to Santa Rosalia, or why are there so many kinds of animals? American Naturalist **93**:145–159.

Jones, C.G., J.H. Lawton, and M. Shachak. 1997. Positive and negative effects of organisms as physical ecosystem engineers. Ecology **78**:1946–1957.

Krebs, C.J., S. Boutin, R. Boonstra, A.R.E. Sinclair, J.N.M. Smith, M.R.T. Dale, K. Martin, and R. Turkington. 1995. Impact of food and predation on the snowshoe hare cycle. Science **269**:1112–1115.

Laska, M.S., and J.T. Wootton. 1998. Theoretical concepts and empirical approaches to measuring interaction strength. Ecology **79**:461–476.

Lewis, J.R. 1964. The ecology of rocky shores, First edition. The English Universities Press Ltd., London.

Lindeman, R.L. 1942. Thc trophic-dynamic aspect of ecology. Ecology **23**:399–418.

Louda, S.M., and S.K. Collinge. 1992. Plant resistance to insect herbivores: a field test of the environmental stress hypothesis. Ecology **73**:153–169.

Lubchenco, J., and B.A. Menge. 1978. Community development and persistence in a low rocky intertidal zone. Ecological Monographs **48**:67–94.

MacArthur, R.H. 1972. Strong, or weak, interactions? Transactions of the Connecticut Academy of Arts and Sciences **44**:177–188.

Marquis, R.J., and C.J. Whelan. 1994. Insectivorous birds increase growth of white oak through consumption of leaf-chewing insects. Ecology **75**:2007–2014.

Martinez, N.D. 1991. Artifacts or attributes? Effects of resolution on the Little Rock Lake food web. Ecological Monographs **61**:367–392.

May, R.M. 1973. Stability and complexity in model ecosystems. Princeton University Press, Princeton, N.J.

McClanahan, T.R. 1992. Epibenthic gastropods of the middle Florida Keys: the role of habitat and environmental stress on assemblage composition. Journal of Experimental Marine Biology and Ecology **160**:169–190.

Melian, C.J., and J. Bascompte. 2004. Food web cohesion. Ecology **85**:352–358.

Menge, B.A. 1976. Organization of the New England rocky intertidal community: role of predation, competition and environmental heterogeneity. Ecological Monographs **46**:355–393.

Menge, B.A. 1992. Community regulation: under what conditions are bottom-up factors important on rocky shores? Ecology **73**:755–765.

Menge, B.A. 1995. Indirect effects in marine rocky intertidal interactions webs: patterns and importance. Ecological Monographs **65**:21–74.

Menge, B.A. 2000. Top-down and bottom-up community regulation in marine rocky intertidal habitats. Journal of Experimental Marine Biology and Ecology **250**:257–289.

Menge, B.A. 2003a. Bottom-up:top-down determination of rocky intertidal shorescape dynamics. Pages 62–81 *in* G. A. Polis, M.E. Power, and G. Huxel, editors, Food webs at the landscape level. University of Chicago Press, Chicago, Ill.

Menge, B.A. 2003b. The overriding importance of environmental context in determining the outcome of species deletion experiments. Pages 16–43 *in* P. Karieva and S.A. Levin, editors, The importance of species: perspectives on expendability and triage. Princeton University Press, Princeton, N.J.

Menge, B.A., E.L. Berlow, C.A. Blanchette, S.A. Navarrete, and S.B. Yamada. 1994. The keystone species concept: variation in interaction strength in a rocky intertidal habitat. Ecological Monographs **64**:249–286.

Menge, B.A., C.A. Blanchette, P. Raimondi, T.L. Freidenburg, S.D. Gaines, J. Lubchenco, D. Lohse, G. Hudson, M.M. Foley, and J. Pamplin. 2004. Species interaction strength: testing model predictions along an upwelling gradient. Ecological Monographs **74**:663–684.

Menge, B.A., B.A. Daley, and P.A. Wheeler. 1996. Control of interaction strength in marine benthic communities. Pages 258–274 *in* G. A. Polis and K. O. Winemiller, editors, Food webs: integration of pattern and dynamics. Chapman and Hall, N.Y.

Menge, B.A., B.A. Daley, P.A. Wheeler., E. Dahlhoff, E. Sanford, and P.T. Strub. 1997. Benthic-pelagic links and rocky intertidal communities: bottom-up effects on top-down control? Proceedings of the National Academy of Sciences USA **94**:14530–14535.

Menge, B.A., and T.M. Farrell. 1989. Community structure and interaction webs in shallow marine hard-bottom communities: tests of an environmental stress model. Advances in Ecological Research **19**:189–262.

Menge, B.A., and T.L. Freidenburg. 2001. Keystone species. Pages 613–631 *in* S.A. Levin, editor, Encyclopedia of biodiversity. Academic Press, N.Y.

Menge, B.A., J. Lubchenco, M.E.S. Bracken, F. Chan, M.M. Foley, T.L. Freidenburg, S.D. Gaines, G. Hudson, C. Krenz, H. Leslie, D.N.L. Menge, R. Russell, and M.S. Webster. 2003. Coastal oceanography set the pace of rocky intertidal community dynamics. Proceedings of the National Academy of Science USA **100**:12229–12234.

Menge, B.A., and A.M. Olson. 1990. Role of scale and environmental factors in regulation of community structure. Trends in Ecology and Evolution **5**:52–57.

Menge, B.A., A.M. Olson., and E.P. Dahlhoff. 2002a. Environmental stress, bottom-up effects, and community dynamics: integrating molecular-physiological and ecological approaches. Integrative and Comparative Biology **42**:892–908.

Menge, B.A., E. Sanford, B.A. Daley, T.L. Freidenburg, G. Hudson, and J. Lubchenco. 2002b. An inter-hemispheric comparison of bottom-up effects on community structure: insights revealed using the comparative-experimental approach. Ecological Research **17**:1–16.

Menge, B.A., and J.P. Sutherland. 1976. Species diversity gradients: synthesis of the roles of predation, competition, and temporal heterogeneity. American Naturalist **110**:351–369.

Menge, B.A., and J.P. Sutherland. 1987. Community regulation: variation in disturbance, competition, and predation in relation to environmental stress and recruitment. American Naturalist **130**:730–757.

Mills, L.S., M.E. Soule, and D.F. Doak. 1993. The keystone-species concept in ecology and conservation. BioScience **43**:219–224.

Morgan, S.G. 2001. The larval ecology of marine communities. Pages 159–181 *in* M.D. Bertness, S.D. Gaines, and M.E. Hay, editors, Marine community ecology. Sinauer Associates, Sunderland, Mass.

Murdoch, W.W. 1966. "Community structure, population control, and competition"—A critique. American Naturalist **100**:219–226.

Navarrete, S.A., B.R. Broitman, E.A. Wieters, and J.C. Castilla. 2005. Scales of benthic-pelagic coupling and the intensity of species interactions: from recruitment limitation to top down control. Proceedings of the National Academy of Science USA **102**:18046–18051.

Navarrete, S.A., and B.A. Menge. 1996. Keystone predation and interaction strength: interactive effects of predators on their main prey. Ecological Monographs **66**:409–429.

Nielsen, K.J. 2001. Bottom-up and top-down forces in tidepools: test of a food chain model in an intertidal community. Ecological Monographs **71**:187–217.

Nielsen, K.J. 2003. Nutrient loading and consumers: agents of change in open-coast macrophyte assemblages. Proceedings of the National Academy of Science USA **100**:7660–7665.

Nielsen, K.J. and S.A. Navarrete. 2004. Mesoscale regulation comes from the bottom-up: intertidal interactions between consumers and upwelling. Ecology Letters **7**:31–41.

Oksanen, L., S.D. Fretwell, J. Arruda, and P. Niemela. 1981. Exploitation ecosystems in gradients of primary productivity. American Naturalist **118**:240–261.

Olson, A.M. 1992. Evolutionary and ecological interactions affecting seaweeds. PhD Thesis. Oregon State University, Corvallis, Oreg.

Paine, R.T. 1966. Food web complexity and species diversity. American Naturalist **100**:65–75.

Paine, R.T. 1969. A note on trophic complexity and community stability. American Naturalist **103**:91–93.

Paine, R.T. 1974. Intertidal community structure: experimental studies on the relationship between a dominant competitor and its principal predator. Oecologia (Berlin) **15**:93–120.

Paine, R.T. 1980. Food webs: linkage, interaction strength and community infrastructure. Journal of Animal Ecology **49**:667–685.

Paine, R.T. 1988. Food webs: road maps of interactions or grist for theoretical development? Ecology **69**:1648–1654.

Paine, R.T. 1992. Food-web analysis through field measurement of per capita interaction strength. Nature **355**:73–75.

Paine, R.T. 1995. A conversation on refining the concept of keystone species. Conservation Biology **9**:962–964.

Pimm, S.L. 1982. Food webs. Chapman and Hall, London.

Polis, G.A. 1991. Complex trophic interactions in deserts: an empirical critique of food-web theory. American Naturalist **138**:123–155.

Polis, G.A. 1994. Food webs, trophic cascades and community structure. Australia Journal of Ecology **19**:121–136.

Polis, G.A. 1999. Why are parts of the world green? Multiple factors control productivity and the distribution of biomass. Oikos **86**:3–15.

Polis, G.A., and S.D. Hurd. 1996. Linking marine and terrestrial food webs: allochthonous input from the ocean supports high secondary productivity on small islands and coastal land communities. American Naturalist **147**:396–423.

Polis, G.A., and D. Strong. 1996. Food web complexity and community dynamics. American Naturalist **147**:813–846.

Power, M.E., D. Tilman, J.A. Estes, B.A. Menge, W.J. Bond, L.S. Mills, G. Daily, J.C. Castilla, J. Lubchenco, and R.T. Paine. 1996. Challenges in the quest for keystones. BioScience **46**:609–620.

Raffaelli, D.G., and S.J. Hall. 1996. Assessing the relative importance of trophic links in food webs. Pages 185–191 *in* G.A. Polis and K.O. Winemiller, editors, Food webs: integration of patterns and dynamics. Chapman and Hall, N.Y.

Reuman, D.C., and J.E. Cohen. 2004. Trophic links' length and slope in the Tuesday Lake food web with species's body mass and numerical abundance. Journal of Animal Ecology **73**:852–866.

Roughgarden, J., S.D. Gaines, and H. Possingham. 1988. Recruitment dynamics in complex life cycles. Science **241**:1460–1466.

Sala, E., and M.H. Graham. 2002. Community-wide distribution of predator-prey interaction strength in kelp forests. Proceedings of the National Academy of Science USA **99**:3678–3683.

Schiel, D.R. 2004. The structure and replenishment of rocky shore intertidal communities and biogeographic comparisons. Journal of Experimental Marine Biology and Ecology **300**:309–342.

Schmitt, R.J. 1987. Indirect interactions between prey: apparent competition, predator aggregation, and habitat segregation. Ecology **68**:1887–1897.

Schmitz, O.J. 2003. Top predator control of plant biodiversity and productivity in an old-field ecosystem. Ecology Letters **6**:156–163.

Schmitz, O.J., P.A. Hamback, and A.P. Beckerman. 2000. Trophic cascades in terrestrial systems: a review of the effects of carnivore removals on plants. American Naturalist **155**:141–153.

Schoener, T.W. 1989. Food webs from the small to the large. Ecology **70**:1559–1589.

Shurin, J.B., E.T. Borer, E.W. Seabloom, K. Anderson, C.A. Blanchette, B.R. Broitman, W.E. Cooper, and B.S. Halpern. 2002. A cross-ecosystem comparison of the strength of trophic cascades. Ecology Letters **5**:785–791.

Slobodkin, L.B., F.E. Smith, and N.G. Hairston. 1967. Regulation in terrestrial ecosystems, and the implied balance of nature. American Naturalist **101**:109–124.

Stephenson, T.A., and A. Stephenson. 1972. Life between tidemarks on rocky shores. W.H. Freeman, San Francisco.

Strong, D.R. 1992. Are trophic cascades all wet? Differentiation and donor-control in speciose ecosystems. Ecology **73**:747–754.

Tilman, D. 1982. Resource competition and community structure. Princeton University Press, Princeton, N.J.

Tilman, D., and J. A. Downing. 1994. Biodiversity and stability in grasslands. Nature **367**:363-365.

Vinueza, L.R., G.M. Branch, M.L. Branch, and R.H. Bustamante. 2006. Top-down herbivory and bottom-up El Niño effects on Galapagos rocky-shore communities. Ecological Monographs **76**:111–131.

Whittaker, R.H. 1970. Communities and ecosystems. Collier-Macmillan Limited, London.

Williams, R.J., E.L. Berlow, J.A. Dunne, A.-L. Barabasi, and N.D. Martinez. 2002. Two degress of separation in complex food webs. Proceedings of the National Academy of Science USA **99**:12913–12916.

Witman, J. D., and F. Smith. 2003. Rapid community change at a tropical upwelling sie in the Galapagos Marine Reserve. Biodiversity and Conservation **12**:25–45.

Worm, B., H.K. Lotze, H. Hillebrand, and U. Sommer. 2002. Consumer versus resource control of species diversity and ecosystem functioning. Nature **417**:848–851.

2

Kelp Forest Food Webs in the Aleutian Archipelago

James A. Estes

The Aleutian archipelago, which forms a porous boundary between the North Pacific Ocean and southern Bering Sea (fig. 2.1), was created during the late Eocene or early Oligocene by tectonic uplift and volcanism at the northern margin of the northward-moving Pacific plate (Gard 1978, Avé Lallemant and Oldow 2000). The Fox Islands, eastern-most of the island groups comprising the Aleutian archipelago, lie on the North American continental shelf and as such are land-bridge islands; the remaining islands in the Aleutian archipelago are oceanic.

Climate has changed appreciably over the geological history of the Aleutian Islands. Subtropical conditions prevailed at the time of their genesis but substantial cooling began at the end of the Miocene with the onset of the most recent glacial age (Addicott 1969). Ice cover during Pleistocene glacial advances likely destroyed or greatly diminished the region's terrestrial and shallow-water marine biotas. The modern climate is subarctic maritime (Armstrong 1977). Summer weather, which runs from late May through August, is characteristically calm whereas intense storms are common during the remainder of the year. Surface sea temperatures range between a summer high of about 7°C and a winter low of about 1°C. Prevailing ocean currents are dominated by the westward flowing Alaskan Stream to the south and the eastward flowing Aleutian North Slope Current to the north (Stabeno et al. 1999). Marine waters surrounding the Aleutian Islands are ice-free throughout the year, except along the

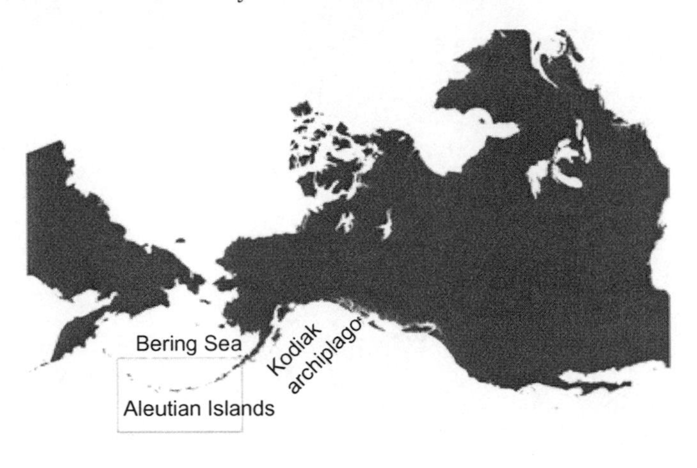

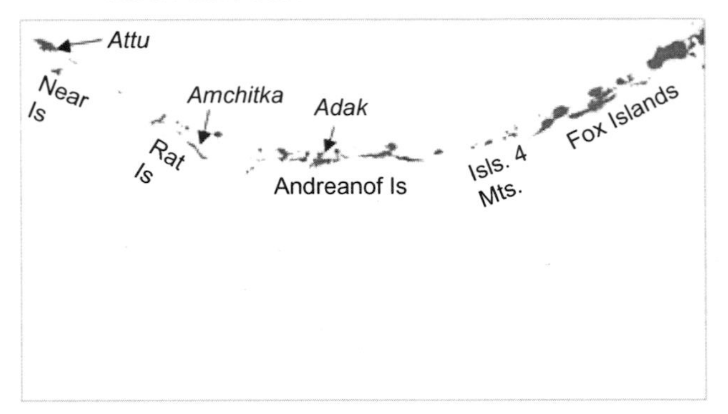

FIGURE 2.1 Map of the North Pacific and Aleutian Islands region showing place names referred to in this chapter.

northern coastlines of the eastern Fox Islands during unusually cold winters. Nutrient-rich seawater and high summer solar radiation result in high primary productivity (Walsh et al. 1989), supporting in turn spectacular numbers of marine birds, mammals, and fishes.

Rocky shorelines surround the Aleutian Islands, often extending to great depths. The absence of a continental shelf combined with modest terrigenous influx result in a paucity of soft-sediment habitats. The absence of a continental shelf also creates substantially different oceanographic and biological conditions from those that occur along continental margins. Long-shore currents in this region may not advect coastal surface waters seaward because they flow in a counter-clockwise direction relative to island landmasses, thus potentially entraining the dispersive life-history stages of coastal marine species within the coastal zone and fueling the growth of adult life stages that might otherwise be recruitment-limited. The expanses of deep-water rocky substratum also provide extensive habitat for various reef-inhabiting species that otherwise might be restricted to smaller population sizes in shallower waters. As explained later in the chapter, these features have powerful influences on coastal food-web dynamics.

The Biological Setting

In contrast with many tropical oceanic archipelagos, the marine biota of the Aleutian archipelago is both highly impoverished and contains few endemic species—qualities that likely resulted from glacially induced extinctions during the Pleistocene and reinvasions from Asia or North America following early Holocene glacial recessions. Species composition is remarkably consistent across the archipelago, a notable exception occurring near the western end of the Fox Islands where species like bull kelp (*Nereocystis leutkeana*) and the sunflower star (*Pycnopodia helianthiodes*) are abundant to the east but rare (*Pycnopodia*) or absent (*Nereocystis*) to the west (Miller and Estes 1989; J.A. Estes, pers. observ.). The distribution and abundance of a few other species or taxa seem to vary across the Aleutian archipelago in relation to their continent of origin. For instance, the Asian kelp (*Thallasiophyllum clathrus*) is abundant from the Near Islands through the Islands of Four Mountains, comparatively rare in the Fox Islands, and absent in the Kodiak archipelago (B. Konar, pers. commun.). Hexagrammid fishes (a common kelp forest taxon in the Aleutians) vary in relative abundance across the archipelago, with rock greenling (*Hexagrammus lagocephalus*) and kelp greenling (*H. decagrammus*) being the numerically dominant forms in the west and the east, respectively.

The Kelp Forest Ecosystem

Kelp forest communities, which characterize shallow rocky sublittoral habitats in the Aleutian Islands, support a diverse functional array of consumers including macro and meso herbivores, carnivores, detritovores, filter feeders, and suspension feeders. Kelps and other brown algae typically occur from the sublittoral fringe to depths of 20–25 m. Red algae range somewhat deeper and the distributions of many kelp forest heterotrophs, such as sea urchins, extend far deeper still. Kelp forest plants form four distinct canopy layers. These include a "pavement" of crustose coralline algae dominated by a single species, *Clathromorphum nereostratum*. A "turf canopy" of red fleshy and articulated coralline algae extends 20–30 cm above the coralline pavement. Kelps and a variety of other fleshy frondose algae form an "epibenthic canopy" a meter or more above the seafloor. The epibenthic canopy is dominated by several species of *Laminaria* in shallow water, and by *Agarum cribrosum* in deeper water (Dayton 1975, Estes et al. 1978). *Laminaria* is competitively dominant over *Agarum* (Dayton 1975) whereas *Agarum*, apparently by virtue of its higher concentration of secondary metabolites (Estes and Steinberg 1988), is more resistant than *Laminaria* to herbivory. Most epibenthic canopy kelps are perennials. The annual kelp, *Alaria fistulosa*, is a seasonal species that grows rapidly during late spring and early summer, usually reaching the sea-surface by early May. *Alaria fistulosa* senesces rapidly in late summer and the surface canopy has largely disappeared by late September or early October. *Alaria fistulosa*, a competitive subordinate to both *Laminaria* and *Agarum* (Dayton 1975), commonly grows in disturbed habitats with high wave energy, unconsolidated substrates, or high intensities of grazing.

Kelp Forest Food Webs

Approach to study Food webs provide a useful template for thinking about the connectivity among species and the dynamics of population and ecosystem change (Paine 1980). The pathways by which species can interact with one another within a food web are enormously varied and complex. For instance, an assemblage of just 10 species contains 9,864,090 potential food-web pathways (Estes et al. 2004). While some of these pathways cannot or simply do not occur (e.g., plants seldom eat carnivores), this hypothetical example is sufficient to demonstrate that the list of species in an ecosystem is merely the starting point for appreciating functional biodiversity. Understanding the complex network of interaction pathways and their dynamics is presently an unrealistic goal for any natural ecosystem. A more sensible approach requires three simplifying actions. One is a focus on some smaller subset of species and food-web pathways. Here, the key to success is selecting those that matter. Another is to develop a conceptual foundation for asking questions, such as sorting out the relative importance of bottom-up versus top-down forcing processes (Hunter and Price 1992). Finally, because the dynamics of static systems are difficult to understand (May 1973), in order to properly gauge the workings of any food-web pathway, some part of that pathway (usually a single species) must be perturbed.

The following account of kelp forest food webs in the Aleutian Islands is based on these simplifying approaches. Our understanding has been aided by several important features of the ecosystem. One is that the region is comprised of islands, each with discrete boundaries and unique histories, thus providing opportunities for ecologically interesting comparisons. Another is that the previously described geologic and climatic uniformity of the islands reduces confounding influences in inter-island comparisons. Yet another is that this region has been perturbed by various human activities in ways that make variation among islands and through time informative to the understanding of food-web dynamics.

My studies have focused on sea otters and several other species that are intimately connected to sea otters through the food web: sea urchins, their principal prey; kelps and other fleshy macroalgae, the main autotrophs and the primary food of sea urchins; and a variety of consumers whose natural histories are linked to kelp forests in various ways. The primary perturbation around which most planned analyses were based was the waxing and waning of sea otter populations: their near-extinction and subsequent recovery from the Pacific maritime fur trade, which ended in 1911 (Kenyon 1969), and a second precipitous decline during the 1990s (Doroff et al. 2003) purportedly caused by increased killer whale predation (Estes et al. 1998). Disturbances to the surrounding oceanic ecosystem—whaling, fishing, and ocean regime shifts—provide a second set of perturbations around which a variety of unplanned analyses were conducted. The conceptual bases for all of these analyses were founded on the following questions:

1. What is the relative importance of bottom-up and top-down forcing processes?
2. Do trophic cascades (*sensu* Paine 1980, Carpenter and Kitchell 1993) occur, and if so, does plant/herbivore interaction strength alternate predictably between odd and even length food chains, as predicted by Fretwell (1987)?
3. What is the nature and importance of indirect food-web effects?
4. To what extent can identified processes be generalized in space and time?
5. To what extent are the food-web dynamics connected over large scales of time and space—i.e., with historically important events or with events in the open sea?
6. What are the evolutionary consequences of food-web interactions to the behavior and life history of donor and recipient species?

Methods The Pacific maritime fur trade reduced sea otter numbers in the Aleutian archipelago from a hundred thousand or more in the mid 1700s to less than a thousand in the early 1900s (Kenyon 1969, Estes 1977, Doroff et al. 2003). The surviving animals occurred in two or three remnant colonies. Although otter numbers increased rapidly following the cessation of harvest (Estes 1990, Bodkin et al. 1999), range spread occurred slowly from the remnant colonies so that the overall recovery pattern was spatially and temporally asynchronous. For instance, several islands in the Rat Island group (fig. 2.1) had recovered to pre-exploitation levels by the late 1930s or early 1940s whereas other islands in the Near and Four Mountains groups remained uninhabited by sea otters into the 1970s and 1980s. In the late 1980s or early 1990s, sea otter numbers in the Aleutian Islands turned rapidly downward, decreasing from about 75,000 in 1990 to about 6,000 by 2000 (Doroff et al. 2003). The sea otter's role in kelp forest food-web dynamics has thus been inferred from differences in the abundance and behavior of various species between islands with and without sea otters, and through time as otter numbers increased or declined. Stable carbon isotope analyses have been used to determine how the sources of primary production differ between systems with and without sea otters.

Food-Web Dynamics

Direct effects of sea otter predation Inter-island comparisons indicate that populations of various benthic macroinvertebrates are sharply limited by sea otter predation. For example, the standing biomass of sea urchins in the Aleutian Islands is roughly an order of magnitude less where sea otters are abundant than where sea otters are absent (Estes and Duggins 1995). Sea otters selectively consume the larger sea urchins and thus these individuals (≈40–100 mm test diameter) are missing from populations that are exploited by sea otters (Estes et al. 1978, Estes and Duggins 1995). As air-breathing predators, sea otter foraging efficiency declines with water depth, an effect that is apparent in the differing depth distributions of sea urchins at islands with and without sea otters. Where otters are absent, sea urchins are most abundant at the sublittoral fringe and their density declines with depth; where sea otters are abundant, sea urchins are relatively rare at the sublittoral fringe and their density increases with depth (Dayton 1975, Estes et al. 1978, Estes and Steinberg 1988).

The sea otter population at Amchitka Island existed at or near equilibrium density (about 6,500 individuals—Estes 1977) from at least the 1960s through the 1980s. The distribution, abundance, and population structure of sea urchins also remained largely unchanged during this period (Estes and Duggins 1995, Watt et al. 2000), implying a demographic balance between the gains from recruitment and growth on the one hand, and losses to sea otter predation on the other. The feasibility of these purported balancing processes can be assessed by contrasting an estimate of sea urchin production (based on habitat area and sea urchin density, population structure, and size-specific growth rate) with an estimate of losses to sea otter predation. Watt and colleagues (2000) derived the latter values from estimates of otter abundance and foraging range, per capita consumption rate and size selectivity as determined by Estes and Duggins (1995). Sea otters in the Aleutian Islands rarely dive beyond a depth of 10 m to forage on sea urchins (Watt et al. 2000). When the 10-m depth contour is thus used to delineate sea otter foraging habitat, the resulting estimated loss of sea urchins to sea otter predation is grossly unsustainable. Accordingly, the equilibrium dynamics of this predator–prey system can only be maintained through prey immigration, which must occur, as urchins from deeper water are attracted to the shallows by nutritional rewards provided by kelps and other fleshy algae. Predictable and strong recruitment together with the vast areas of deep-water rocky habitat that surround the Aleutian Islands potentially provide a source population of sufficient size and productivity to maintain the predator–prey equilibrium (Estes and Duggins 1995). This deep-water subsidy and production potential might explain the remarkably high sea otter population densities (≈40 individuals per km of shoreline) reported from the Aleutian Islands (Estes 1977) as well as the extraordinary speed with which kelp forests in the central and western Aleutians became urchin barrens following sea otter declines in the 1990s (Estes et al. 1998, 2004).

Trophic cascades Trophic cascades occur when top-down effects interconnect species of successively lower trophic status across multiple trophic levels (Paine 1980). Such interactions initiated by apex predators commonly influence herbivore/plant interactions at the base of the food web (Carpenter and Kitchell 1993, Pace et al. 1999). Some of the earliest evidence for trophic cascades came from studies of kelp forest ecosystems in the Aleutian Islands. As described in the preceding section, sea urchins were found to be abundant at islands where otters were rare or absent and relatively rare at islands where otters were abundant. Sea urchins eat kelp and other fleshy algae, and thus the abundance and distribution of these species also vary substantially between islands with and without sea otters. That is, islands with abundant otters also support dense kelp forests whereas islands lacking sea otters are typically characterized by sea urchin barrens (fig. 2.2). Once established, these community states resist change to the alternate state (Konar and Estes 2003).

Generality of trophic cascades While few would quibble over the existence of the otter–urchin–kelp trophic cascade, some have questioned whether it is rare or common (Foster and Schiel 1988). This question has been addressed in the Aleutian Islands through more extensive “representative” sampling of islands with and without sea otters. This was accomplished by measuring kelp and urchin abundances within randomly placed quadrats on the seafloor at numerous

FIGURE 2.2 Alternate phase states of Aleutian Islands kelp forest ecosystems: (a) kelp-dominated (photo courtesy Paul Dayton); (b) sea urchin barren (photo courtesy Michael Kenner).

randomly selected points along the shorelines of the various islands. The resulting data established that while kelp densities varied considerably at otter-dominated islands and urchin densities varied to a similar degree at otter-free islands, rocky-reef communities occurred as either kelp forests or urchin barrens, with intermediates seldom found, and these phase-states were broadly predictable based on the presence or absence of sea otters (Estes and Duggins 1995, fig. 2.3).

Indirect effects of sea otter predation The otter–urchin–kelp trophic cascade influences numerous other species and food-web processes. These indirect effects appear to occur in three general ways: as

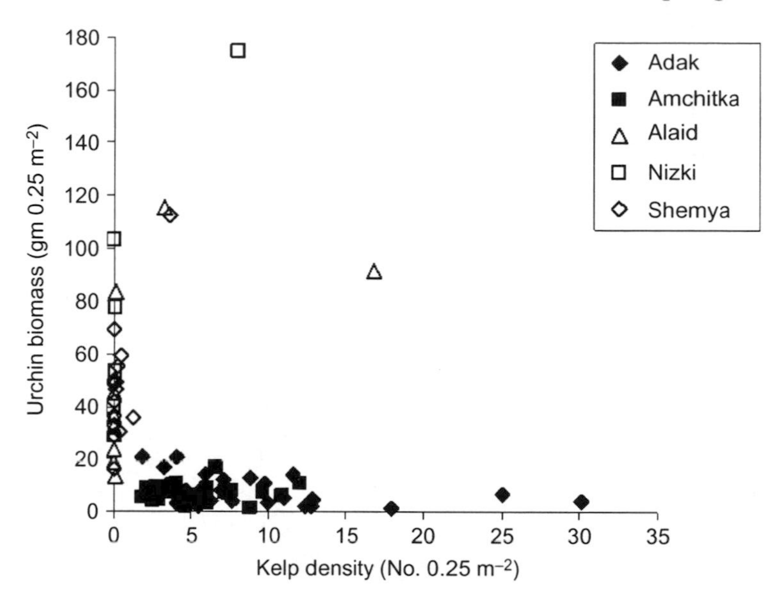

FIGURE 2.3 Phase diagram of sea urchin biomass vs. kelp density at islands with and without sea otters in the Aleutian archipelago. Reproduced from Estes and Duggins 1995.

habitat effects associated with the presence or absence of kelp forests; as production effects associated with exceedingly high rates of kelp photosynthesis and growth; and as prey availability effects on higher trophic forms that result directly or indirectly (sometimes in serpentine ways) from sea otter predation. The vast majority of potential food-web pathways leading outward from sea otter predation have not even been imagined, much less explored. Table 2.1 provides a synopsis of those that have been studied.

Evolutionary influences of trophic cascades Strong species interactions should lead to selection and evolutionary change, an expectation that has been explored for the sea otter/kelp forest system by focusing on defense and resistance in the macroalgae and their principal herbivores. This focus was taken for two reasons. First, defense and resistance between plants and their herbivores was well known in other species and systems (Rosenthal and Berenbaum 1992). Second, earlier ecological studies led to the expectation that plant/herbivore interactions would evolve in fundamentally different ways depending upon whether sea otters, or similar predators, were present or absent in an ecosystem over a sufficiently long time. The rationale for this expectation followed from the well-known model of food-chain dynamics (fig. 2.4) developed by Hairston and colleagues (1960) and refined by Fretwell (1987). This model specifically predicts that autotrophs in even-numbered food chains are under strong selection to

Table 2.1. Summary of indirect food-web effects by sea otters on other coastal marine species in the Aleutian archipelago.

Species or Process	Sea Otters Abundant	Sea Otters Rare or Absent	Responsible Mechanism	Source
Primary production	Comparatively high	About 3–4 times lower	Presence of abundant kelp because of otter–urchin–kelp trophic cascade; high rate of organic carbon fixation by kelp photosynthesis	Duggins et al. (1989); Simenstad et al. (1993)
Filter feeding invertebrates	Growth rates high	Growth rates comparatively low	Increased primary production from otter–urchin–kelp trophic cascade;	Duggins et al. (1989)
Kelp forest fishes	Abundant; fauna numerically dominated by kelp greenling	Relatively rare; fauna less dominated by single species	Habitat, production, and prey community differences associated with otter–urchin–kelp trophic cascade	Reisewitz (2006)
Glaucous winged gulls	Mainly piscivorous	Feeds mainly on intertidal invertebrates	Differences in fish abundance resulting from otter–urchin–kelp trophic cascade; reduction of intertidal invertebrates by sea otter predation	Irons et al. (1986)
Common eiders	Rare	Common	Competition with sea otters for benthic invertebrate prey resources	D.B. Irons, G.V. Byrd, and J.A. Estes, unpubl. data

Sea stars	Small and rare	Large and common	Direct predation by sea otters; possibly inhibition of stars by benthic algae resulting from otter–urchin–kelp trophic cascade.	Vicknair and Estes, unpublished data
Sea star predation on mussels	Low	Comparatively high	Direct and indirect effects of sea otters on sea star populations	Vicknair and Estes, unpublished data
Bald eagles	Kelp forest fishes and sea otter common in diet	Kelp forest fishes and otter pups rare; seabirds and Atka Mackerel more common	Availability differences in sea otters; effects of otter–urchin–kelp trophic cascade on kelp forest fishes; presumed dietary switching to less beneficial prey	R.G. Anthony, unpubl. data
Macroalgae	Interspecies competition intense	Interspecies competition weak or absent	Abundant macroalgae resulting from otter–urchin–kelp trophic cascade; space and light limitation	Dayton (1975)
Sea otter populations	Carrying capacity elevated	Carrying capacity lower	Positive feedback loop of otter–urchin–kelp trophic cascade on otter numbers through increased production and dietary expansion to include kelp forest fishes	Estes et al. (1978); Estes (1990)

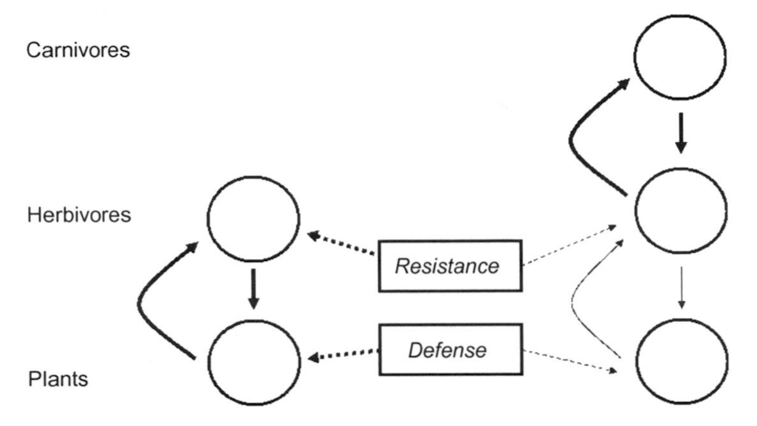

FIGURE 2.4 Hypothesized evolutionary consequences of food-chain length in systems under top-down control. Strong plant–herbivore interactions occur in systems with two (or any even number) trophic levels. This results in strong (indicated by heavy lines) selection for the coevolution of plant defenses against herbivory and herbivore resistance to those defenses. Addition of a third trophic level (or any odd number of trophic levels) weakens (indicated by lighter lines) the influence of herbivores on plants, thus reducing selection for the coevolution of defense and resistance (after Fretwell 1987 and Steinberg et al. 1995). From Estes and colleagues (2004).

defend themselves against herbivores whereas reduced herbivory in odd numbered food chains leads to more poorly defended plants and thus a reduced selection intensity for resistance to plant defenses by the herbivores.

This hypothetical evolutionary model predicts that kelp forest plants in the North Pacific Ocean are poorly defended, and that their herbivores are unable to resist these defenses. Contrasting brown algal phlorotannin concentrations and herbivore resistance to phlorotannins between northeastern and southwestern Pacific kelp forests tested these predictions. The southwest Pacific was chosen as a point of contrast because, lacking predators of comparable influence to the sea otter (Estes and Steinberg 1988), that region was viewed as a two-trophic level system. Phlorotannins were chosen as the purported mode of defense because marine algae commonly use secondary metabolites to deter their herbivores (Hay and Fenical 1988) and phlorotannins are the most likely candidate for this role in marine brown algae (Steinberg 1992). Studies found that phlorotannin concentrations are roughly an order of magnitude greater in southern than northern hemisphere kelps and rockweeds. Additionally, feeding trials, in which herbivorous snails and sea urchins were offered palatable foods with varying sources and concentrations of phlorotannins, demonstrated that northern hemisphere herbivores are more strongly deterred by these compounds than are their southern hemisphere counterparts (Steinberg et al. 1995).

These various findings provide support for the hypothetical evolutionary model (fig. 2.4). Specifically, they suggest that sea otters and their recent ancestors decoupled an evolutionary arms race that otherwise might have led to very different life history patterns in the kelps and their herbivores through the coevolution of defense and resistance. This interpretation may explain why northern hemisphere kelp forests are so extensively deforested by overgrazing in regions where the apex predators have been lost (Harrold and Pearse 1987, Steneck et al. 2003).

Oceanic prey subsidies During most years, sea otters in the Aleutian Islands feed on kelp-forest dependent species. The immigration of sea urchins into shallow coastal waters, described above, is an exception to this generalization. Another exception is the inshore migration of smooth lumpsuckers (*Aptocyclus ventricosus*), a slow-moving oceanic fish whose distribution and movements in the Aleutian region change inter-annually for unknown reasons (Yoshida and Yamaguchi 1985, Il'inskii and Radchenko 1992). On occasion, immense numbers of these animals move into the coastal zone and spawn. They appear suddenly in late November or early December and are gone from the coastal zone by late April or early May. These inshore spawning migrations are episodic. Only two such events have been chronicled in the Aleutian Islands, one in the mid 1960s (Kenyon 1969) and the other in the early 1990s (Watt et al. 2000). The latter event ranged widely (Watt et al. 2000), extending eastward to at least Adak Island and westward to at least Attu Island (see fig. 2.1). Winter is normally the time of food-limitation and starvation-induced mortality for sea otters in Alaska (Kenyon 1969, Bodkin et al. 2000). Adult lumpsuckers, when present, thus provide a food subsidy from the oceanic ecosystem that benefits sea otters during this critical period. These benefits are manifested in shorter-duration foraging bouts; reduced overall time spent foraging (Gelatt et al. 2002), improved body condition (Monson et al. 2000), and reduced winter mortality (Watt et al. 2000, Monson et al. 2000).

Aboriginal humans Expansion of modern humans into the New World at the end of the Holocene is thought to have exterminated close to 50 percent of the terrestrial megafauna (Martin 1973, Alroy 2001). Although these early people had well-developed maritime hunting technologies, especially at higher latitudes, their influences on marine mammals and other large marine vertebrates remains poorly known. One exception is the extinct Steller's sea cow (*Hydrodamalis gigas*), which ranged widely across the North Pacific Ocean and southern Bering Sea through the Pleistocene but disappeared from most of this region before the arrival of Europeans. An abundant sea cow population survived in the Commander Islands, apparently the only island group in the Aleutian archipelago that was never peopled. These observations strongly implicate aboriginal human hunting as the principal cause of the sea cow's demise (Domning 1978). Steller's sea cows were exclusively algivorous, feeding on kelps and other fleshy macroalgae in the shallow coastal zone (Steller 1751, Domning 1978). Although the sea cow's role in the kelp forest remains speculative, the extinction of this large and abundant species probably altered the dynamics of kelp forest food webs (Estes et al. 1989).

There is also evidence that aboriginal Aleuts hunted sea otters to low numbers, in turn causing an ecosystem phase shift from kelp forests to sea urchin barrens (Simenstad et al. 1978). In this case, the evidence comes from a combination of patterns seen in extant ecological communities and prehistoric midden remains. As explained above, the size distribution of sea urchins provides a clear proxy for the presence or absence of sea otters. When sea otters are absent, urchin test diameters range upward to 100 mm. When sea otters are present, even in relatively low numbers, the maximum sea urchin test diameter is only about 35 mm (Estes and Duggins 1995). Thus, it is possible to infer whether or not sea otters were present or absent during prehistoric times by measuring the size distributions of their remains in Aleut kitchen middens. Radiocarbon dating indicates that Aleuts occupied Amchitka Island, the site where these analyses were conducted, from about 2,500 BP until shortly after European arrival. Sea urchin test remains dominate Aleut kitchen middens throughout this period. Urchin size frequency analyses by time stratum (fig. 2.5) clearly indicate that the kelp forest system contained otters when Aleuts first occupied this site, but that otters were rare or absent thereafter. These purported effects of Aleut hunting on kelp forest ecosystems presumably were not widespread, given the large numbers of sea otters harvested by the earliest commercial fur hunters (Kenyon 1969).

Ecological chain reactions and the effects of modern industrial exploitation By the mid 1990s it had become clear that sea otter numbers at Adak Island (fig. 2.1) were in significant decline. This unexpected event prompted three immediate questions. How widespread was the decline, what caused it, and what were its influences on kelp forest food-web dynamics? More recent surveys show that sea otter populations have declined throughout the Aleutian archipelago (Doroff et al. 2003), and indeed eastward along the Alaska Peninsula to about the Kodiak Archipelago (Burn and Doroff 2005). The overall rate of population decline was about 18% yr-1, and by 2000, sea otter numbers in the Aleutian Islands had declined by roughly an order of magnitude (Doroff et al. 2003), some 15–20 fold below the estimated equilibrium density (Burn et al. 2003).

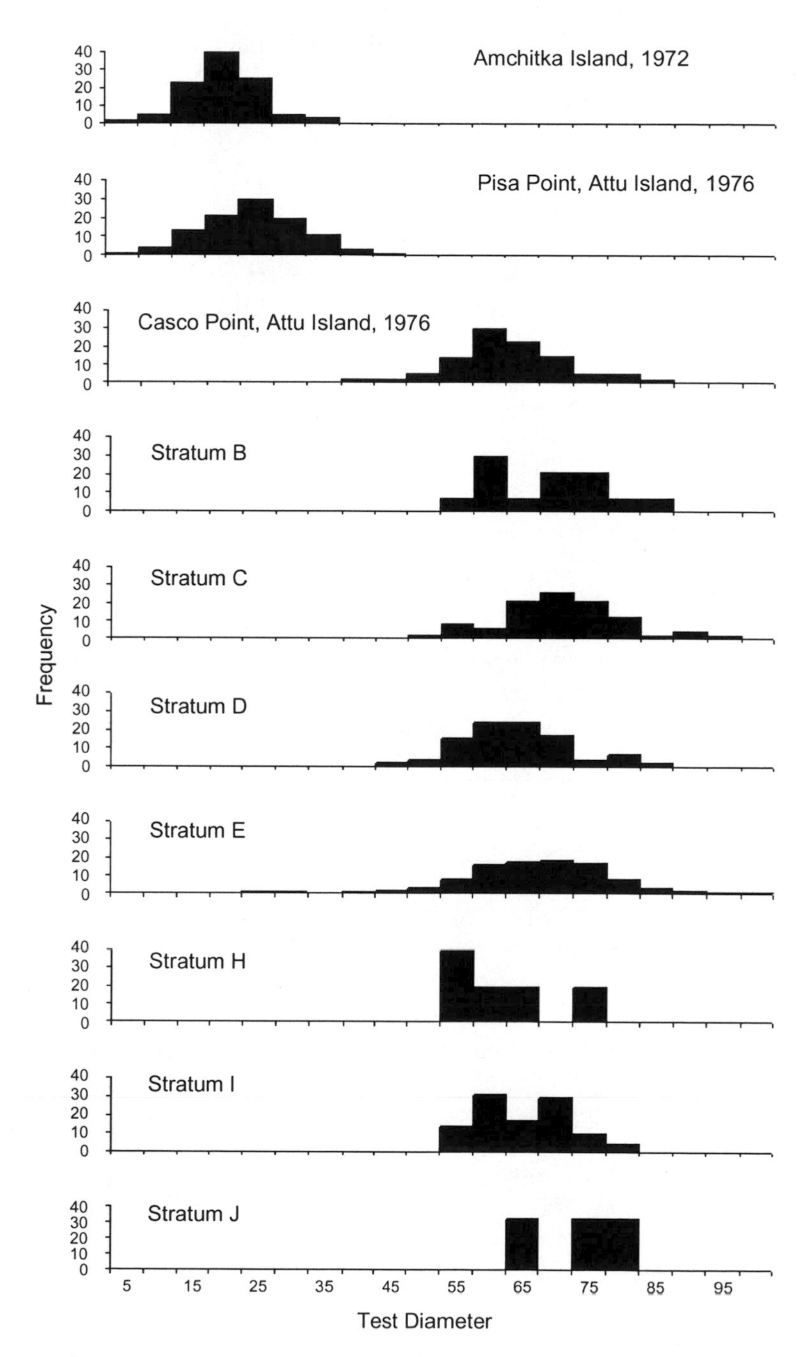

FIGURE 2.5 The size frequency distributions of sea urchins from kelp forest systems in the Aleutian archipelago. Sea otters present (Amchitka; Pisa Pt., Attu); sea otters absent (Casco Point, Attu); midden remains (all others). These patterns suggest that sea otters were absent for nearly the entire period of Aleut occupation. From Simenstad and colleagues (1978).

Ecological consequences of the declining sea otters have been rapid and profound. By the late 1990s subtidal reef systems at Adak and Amchitka islands had completely transitioned to urchin barrens. Overall, sea urchin biomass increased by about eightfold, kelp density declined about 12-fold, and daily grazing intensity increased from near zero to about 50% (Estes et al. 2004). Indirect effects of these changes have been documented in the abundance and species composition of kelp forest fishes (Reisewitz et al. 2006), the diet of bald eagles (R.G. Anthony, unpublished data), and the diet and foraging behavior of glaucous winged gulls (J.A. Estes, unpublished data).

Reasons for the sea otter decline have been more difficult to determine. The weight of current evidence implicates increased killer whale predation (Estes et al. 1998), although the explanation for why this happened is much less clear. The facts that killer whales prey on many other marine mammal species, and that the sea otter decline began on the heels of a similarly abrupt decline in Steller sea lion (*Eumetopias jubata*) numbers, led Estes and colleagues (1998) to speculate that the otter declines were caused by an expanding killer whale diet as their preferred or at least more traditional prey became rare (Estes et al. 1998). The sea lion declines are believed by many to have been driven by nutritional limitation, ultimately resulting from competition with fisheries or an ocean regime shift. While these bottom-up forcing mechanisms are logical expectations, they have little supporting evidence and even appear inconsistent with information from more recent periods of the decline (Anonymous 2002, National Research Council 2003). The sea lion and sea otter declines are part of a wholesale megafaunal collapse in the North Pacific Ocean and southern Bering Sea, which also includes harbor seals (*Phoca vitulina*) and northern fur seals (*Callorhinus ursinus*). Based on the sequential nature of these declines and a re-evaluation of various historical information and current evidence, Springer and colleagues (2003) proposed that they were ultimately driven to a large degree by post-World War II industrial whaling. These authors contend that the great whales were an important food resource for killer whales, and that the great whales' demise triggered an ecological chain reaction when the killer whales turned elsewhere for sustenance—first to harbor seals and northern fur seals, then to sea lions, and finally to sea otters as these prey were progressively depleted (fig. 2.6). Although many of the historical details and earlier processes are still being debated, there can be little doubt that the larger North Pacific marine ecosystem is deeply connected in space and time, and thus that changes in the oceanic ecosystem have led to a fundamental reshaping of kelp forest food-web dynamics.

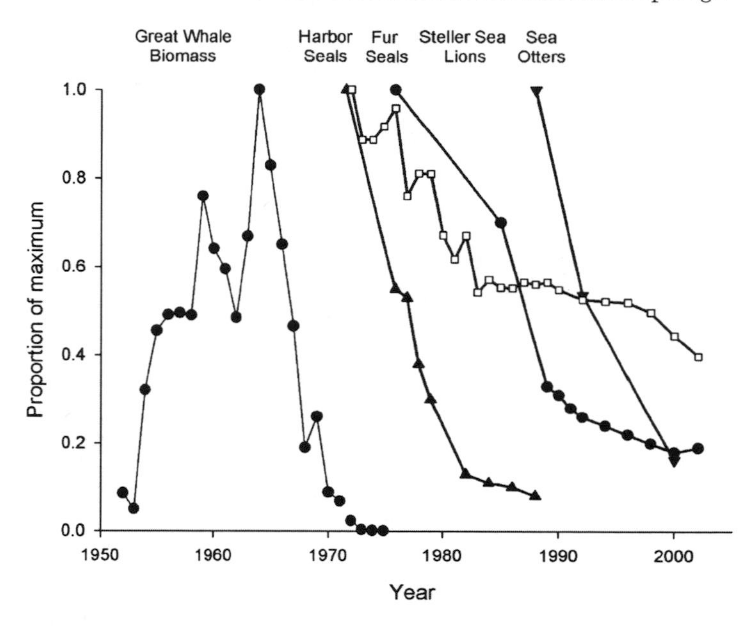

FIGURE 2.6 Reported great whale landings (from International Whaling Commission statistics) and trends in abundance of pinnipeds and sea otters from the Aleutian archipelago and nearby regions of southwest Alaska. Figure modified from Springer and colleagues (2003).

Summary

Aleutian Islands kelp forests are organized around top-down forcing effects and trophic cascades emanating from sea otters, the system's dominant apex carnivore. Studies of food-web dynamics in this system have been built around contrasts in space and time resulting from a large-scale anthropogenic experiment, the over hunting and recovery of sea otters. Sea otters influence the structure and organization of this ecosystem by preying on herbivorous sea urchins, in turn releasing kelps and other benthic algae from limitation by herbivory. This trophic cascade has a wide range of indirect effects on other species and ecological processes. Acting on evolutionary time scales, it seems to have lessened the potential for an arms race between plants and their herbivores, thus creating a biota in which the plants are poorly defended against herbivory and the herbivores are unable to resist plant chemical defenses. Early humans influenced this system by exterminating Steller's sea cow and over-hunting sea otters. More recently, the overexploitation of oceanic resources, especially the large whales, appears to have set off an ecological chain reaction that has led to the decline of sea otters and a collapse of the kelp forest ecosystem. These various findings convey the image of a food web in which the influences of predation are profound, far reaching, and extensively connected in space and time.

References

Addicott, W.O. 1969. Tertiary climatic change in the marginal northeastern Pacific Ocean. Science **165**:583–586.

Alroy, J. 2001. A multispecies overkill simulation of the end-Pleistocene megafaunal mass extinction. Science **292**:1893–1896.

Anonymous. 2002. Is it food II? Summary report of workshop held in Seward, Alaska, 30–31 May, 2001. (Alaska Sea Grant, University of Alaska, Fairbanks, Alaska, 2002). www.alaskasealife.org/isitfoodII.pdf

Armstrong, R.H. 1977. Weather and climate. Pages 53–58 *in* M.L. Merritt and R.G. Fuller, editors, The environment of Amchitka Island, Alaska. National Technical Information Service, U.S. Department of Commerce, TID-26712, Springfield, Va.

Avé Lallemant, H.G., and J.S. Oldow. 2000. Active displacement partitioning and arc-parallel extension of the Aleutian volcanic arc based on Global Positioning System geodesy and kinematic analysis. Geology **28**:739–742.

Bodkin, J.L., B.E. Bellachey, M.A. Cronin, and K.T. Scribner. 1999. Population demographics and genetic diversity in remnant and translocated populations of sea otters. Conservation Biology **13**:1378–1385.

Bodkin, J.L., A.M. Burdin, and D.A. Ryazanov. 2000. Age- and sex-specific mortality and population structure in sea otters. Marine Mammal Science **16**:201–219.

Burn, D.M., and A.M. Doroff. 2005. Decline in sea otter (*Enhydra lutris*) populations along the Alaska Peninsula, 1986–2001. Fishery Bulletin **103**:270–279.

Burn, D.M., A.M. Doroff, and M.T. Tinker. 2003. Carry capacity and pre-decline abundance of sea otters (*Enhydra lutris kenyoni*) in the Aleutian Islands. Northwestern Naturalist **84**:145–148.

Carpenter, S.R., and J.F. Kitchell, editors. 1993. The trophic cascade in lakes. Cambridge University Press, N.Y.

Dayton, P.K. 1975. Experimental studies of algal canopy interactions in a sea otter-dominated community at Amchitka Island, Alaska. Fishery Bulletin **73**:230–237.

Domning, D.P. 1978. Sirenian evolution in the North Pacific Ocean. University of California Publications Geological Sciences **118**:1–176.

Doroff, A.M., J.A. Estes, M.T. Tinker, D.M. Burn, and T.J. Evans. 2003. Sea otter population declines in the Aleutian archipelago. Journal of Mammalogy **84**:55–64.

Duggins, D.O., C.A. Simenstad, and J.A. Estes. 1989. Magnification of secondary production by kelp detritus in coastal marine ecosystems. Science **245**:170–173.

Estes, J.A. 1977. Population estimates and feeding behavior of sea otters. Pages 511–526 *in* M.L. Merritt, and R.G. Fuller, editors, The environment of Amchitka Island. National Technical Information Service, U.S. Department of Commerce, TID-26712, Springfield, Va.

Estes, J.A. 1990. Growth and equilibrium in sea otter populations. Journal of Animal Ecology **59**:385–400.

Estes, J.A., E.M. Danner, D.F. Doak, B. Konar, A.M. Springer, P.D. Steinberg, M.T. Tinker, and T.M. Williams. 2004. Complex trophic interactions in kelp forest ecosystems. Bulletin of Marine Science **74**:621–638.

Estes, J.A., and D.O. Duggins. 1995. Sea otters and kelp forests in Alaska: generality and variation in a community ecological paradigm. Ecological Monographs **65**:75–100.

Estes, J.A., D.O. Duggins, and G. Rathbun. 1989. The ecology of extinctions in kelp forest communities. Conservation Biology **3**:252–264.

Estes, J.A., N.S. Smith, and J.F. Palmisano. 1978. Sea otter predation and community organization in the western Aleutian Island, Alaska. Ecology **59**:822–833.

Estes, J.A., and P.D. Steinberg. 1988. Predation, herbivory and kelp evolution. Paleobiology **14**:19–36.

Estes, J.A., M.T. Tinker, T.M. Williams, and D.F. Doak. 1998. Killer whale predation on sea otters linking coastal with oceanic ecosystems. Science **282**:473–476.

Foster, M.S., and D.R. Schiel. 1988. Kelp communities and sea otters: keystone species or just another brick in the wall? Pages 92–115 *in* G.R. VanBlaricom and J.A. Estes, editors, The community ecology of sea otters. Springer-Verlag, Berlin.

Fretwell, S.D. 1987. Food chain dynamics: the central theory of ecology? Oikos **20**:169–185.

Gard, L.M., Jr. 1978. Geologic history. Pages 13–34 *in* M.L. Merritt and R.G. Fuller, editors, The environment of Amchitka Island, Alaska. National Technical Information Service, U.S. Department of Commerce, TID-26712, Springfield, Va.

Gelatt, T.S., D.B. Siniff, and J.A. Estes. 2002. Activity patterns and time budgets of the declining sea otter population at Amchitka Island, Alaska. Journal of Wildlife Management **66**:29–39.

Hairston, N.G., F.E. Smith, and L.B. Slobodkin. 1960. Community structure, population control, and competition. American Naturalist **94**:421–425.

Harrold, C., and J.S. Pearse. 1987. The ecological role of echinoderms in kelp forests. Pages 137–233 *in* M. Jangoux and J.M. Lawrence, editors, Echinoderm studies. A.A. Balkema, Rotterdam.

Hay, M.E., and W. Fenical. 1988. Marine plant-herbivore interactions: the ecology of chemical defense. Annual Review of Ecology and Systematics **19**:111–145.

Hunter, M.D., and P.W. Price. 1992. Playing chutes and ladders: heterogeneity and the relative roles of bottom-up and top-down forces in natural communities. Ecology **73**:724–732.

Il'inskii, E.N., and V.I. Radchenko. 1992. Distribution and migration of smooth lumpsucker in the Bering Sea. Plenium UDC 597.5 (265.51). Translated from Biologiya Morya **3–4**:19–25.

Irons, D.B., R.G. Anthony, and J.A. Estes. 1989. Foraging strategies of Glaucous-Winged Gulls in rocky intertidal communities. Ecology **67**:1460–1474.

Kenyon, K.W. 1969. The sea otter in the eastern Pacific Ocean. North American Fauna **68**:1–352.

Konar, B. and J.A. Estes. 2003. The stability of boundary regions between kelp beds and deforested areas. Ecology **84**:174–185.

Martin, P.S. 1973. The discovery of America. Science **179**:969–974.

May, R.M. 1973. Stability and complexity in model ecosystems. Princeton University Press, Princeton, N.J.

Miller, K.A., and J.A. Estes. 1989. A western range extension for *Nereocystis leutkeana* in the North Pacific. Botanica Marina **32**:535–538.

Monson, D., J.A. Estes, D.B. Siniff, and J.B. Bodkin. 2000. Life history plasticity and population regulation in sea otters. Oikos **90**:457–468.

National Research Council. 2003. The decline of the Steller sea lion in Alaskan waters: untangling food webs and fishing nets. National Academy Press, Washington, D.C. 204 pages.

Pace, M.L., J.J. Cole, S.R. Carpenter, and J.F. Kitchell. 1999. Trophic cascades revealed in diverse ecosystems. Trends in Ecology and Evolution **14**:483–488.

Paine, R.T. 1980. Food webs: linkage, interaction strength, and community infrastructure. Journal of Animal Ecology **49**:667–685.

Reisewitz, S., J.A. Estes, and S.A. Simenstad. 2006. Indirect food web interactions: sea otters and kelp forest fishes in the Aleutian archipelago. Oecologia **146:**623–631.

Rosenthal, J., and M. Berenbaum, editors. 1992. Herbivores: their interaction with secondary metabolites. Evolutionary and ecological processes. Academic Press, San Diego.

Simenstad, C.A., J.A. Estes, and K.W. Kenyon. 1978. Aleuts, sea otters, and alternate stable state communities. Science **200**:403–411.

Simenstad, C.A., D.O. Duggins, and P.D. Quay. 1993. High turnover of inorganic carbon in kelp habitats as a cause of delta 13 Carbon variability in marine foodwebs. Marine Biology **116**:147–160.

Springer, A.M., J.A. Estes, G.B. van Vliet, T.M. Williams, D.F. Doak, E.M. Danner, K.A. Forney, and B. Pfister. 2003. Sequential megafaunal collapse in the North Pacific Ocean: an ongoing legacy of industrial whaling? Proceedings of the National Academy of Science USA **100**:12223–12228.

Stabeno, P.J., J.D. Shumacher, and K. Ohtani. 1999. The physical oceanography of the Bering Sea. Pages 1–28 *in* T.R. Loughlin and K. Ohtani, editors, Dynamics of the Bering Sea: a summary of physical, chemical, and biological characteristics, and a synopsis of research on the Bering Sea. North Pacific Marine Science Organization (PICES), University of Alaska Sea Grant, AK-SG-99–03.

Steinberg, P.D. 1992. Geographical variation in the interaction between marine herbivores and brown algal secondary metabolites. Pages 51–92 *in* V. Paul, editor, Ecological roles for marine secondary metabolites. Comstock Press, Ithaca, N.Y.

Steinberg, P.D., J.A. Estes, and F.C. Winter. 1995. Evolutionary consequences of food chain length in kelp forest communities. Proceedings of the National Academy of Sciences USA **92**:8145–8148.

Steller, G.W. 1751. De bestiis marinis. Pages 289–398. Novi commentarii Academiae Scientarium Impailis Petropolitanae 2. St. Petersburg.

Steneck, R.S., M.H. Graham, B.J. Bourque, D. Corbett, J.M. Erlandson, J.A. Estes, and M.J. Tegner. 2003. Kelp forest ecosystem: biodiversity, stability, resilience and future. Environmental Conservation **29**:436–459.

Walsh, J.J., C.P. McRoy, L.K. Coachman, J.J. Georing, J.J. Nihoul, T.E. Whitledge, T.H. Blackburn, P.L. Parker, C.D.Wirick, P.G. Shuert, J.M. Grebmeier, A.M. Springer, R.D. Tripp, D.A. Hansell, S. Djenidi, E. Deleersnijder, K. Henriksen, B.A. Lund, P. Andersen, F.E. Müller-Karger, and K. Dean. 1989. Carbon and nitrogen cycling within the Bering/Chukchi Seas: Source regions for organic matter affecting AOU demands of the Arctic Ocean. Progress in Oceanography **22**:277–359.

Watt, J., D.B. Siniff, and J.A. Estes. 2000. Interdecadal change in diet and population of sea otters at Amchitka Island, Alaska. Oecologia **124**:289–298.

Yoshida, H., and H. Yamaguchi. 1985. Distribution and feeding habits of the pelagic smooth lumpsucker, *Aptocyclus ventricosus* (Pallas), in the Aleutian basin. Bulletin of Faculty of Fisheries Hokkaido University **36**:200–209.

3

Trophic Interactions in Subtidal Rocky Reefs on the West Coast of South Africa

George M. Branch

Physical Setting

South Africa is marketed to tourists under the slogan "a world in one country," reflecting its diverse geography and peoples. The phrase also aptly captures the range of oceanographic conditions as two contrasting currents dominate the coastline (Field and Griffiths 1991). On the east coast the Agulhas sweeps southward at core speeds of 2 m/s, bringing nutrient-poor waters of 21°C–26°C from the South Equatorial Current. On the west coast the Benguela Current has diametrically opposite characteristics. Sluggish and slow-moving in its northward drift, the Benguela is dominated by the effects of upwelling, which brings cold nutrient-rich waters of 9°C–15°C up to the surface, fuelling very high productivity and underpinning lucrative industrial fisheries (Field and Griffiths 1991). A series of reviews of the Benguela synthesize its nature and evolution (Shannon 1985), chemistry (Chapman and Shannon 1985), plankton (Shannon and Pillar 1987), major fisheries (Crawford et al. 1987), and the coastal zone (Branch and Griffiths 1988).

This chapter focuses on the subtidal reefs on the west coast of South Africa, particularly those inhabited by kelp beds in the southern Benguela, and carries three central messages: (1) the importance of physical conditions, particularly wind-induced upwelling; (2) the dominant roles of kelp, filter feeders, and bacteria in energy flows; and (3) the significance of biological interactions in regulating community structure and dynamics.

In the 1980s kelp beds in the southern Benguela were the subjects of a detailed program that used energy flow as a central theme. More than 200 papers resulted, and are summarized by Branch and Griffiths (1988) and Field and Griffiths (1991). This program generated the foundation for the food web described here, but I go beyond it to incorporate subsequent studies that have emphasized nontrophic interactions and the top-down influences of predators.

Biogeographic Patterns

As a consequence of the contrasting currents, six biogeographic provinces can be recognized around the coast of southern Africa (fig. 3.1; Brown and Jarman 1978, Emanuel et al. 1992, Sink et al. 2005). The Agulhas Current hugs the east coast where it is associated with the subtropical Natal Province. North of this lies the Maputaland Province, while to the south, the Agulhas Current veers away from the coast as the continental shelf widens to form the Agulhas Bank. As a consequence, inshore waters are cooler in the warm temperate Agulhas Province. The west coast is characterized by cool upwelled waters and has recently been divided into two provinces, the southern cool temperate Namaqua Province and the northern cool temperate Namib Province. Farther north, conditions warm again in Angola. Precise limits of the Angola and Maputaland Provinces remain to be determined. Several patterns emerge. First, diversity is markedly lower on the west coast than the east coast. Second, biomass declines from west to east, reflecting levels of productivity (Bustamante et al. 1995a). Third, endemicity is extremely high, particularly in the south, where it reaches 40%–60% in most groups, providing support for the recognition of distinct biogeographic regions (Awad et al. 2002).

Kelp beds are typical of the two cool temperate provinces, but extend eastward to Cape Agulhas (fig. 3.1). Three species form beds, but two are dominant: *Ecklonia maxima*, which forms canopies because of its gas-filled floating bladders, and *Laminaria pallida*, which grows beneath *E. maxima* and extends into deeper water. *Macrocystis angustifolia* forms localized beds in sheltered waters that are confined to the Cape Peninsula.

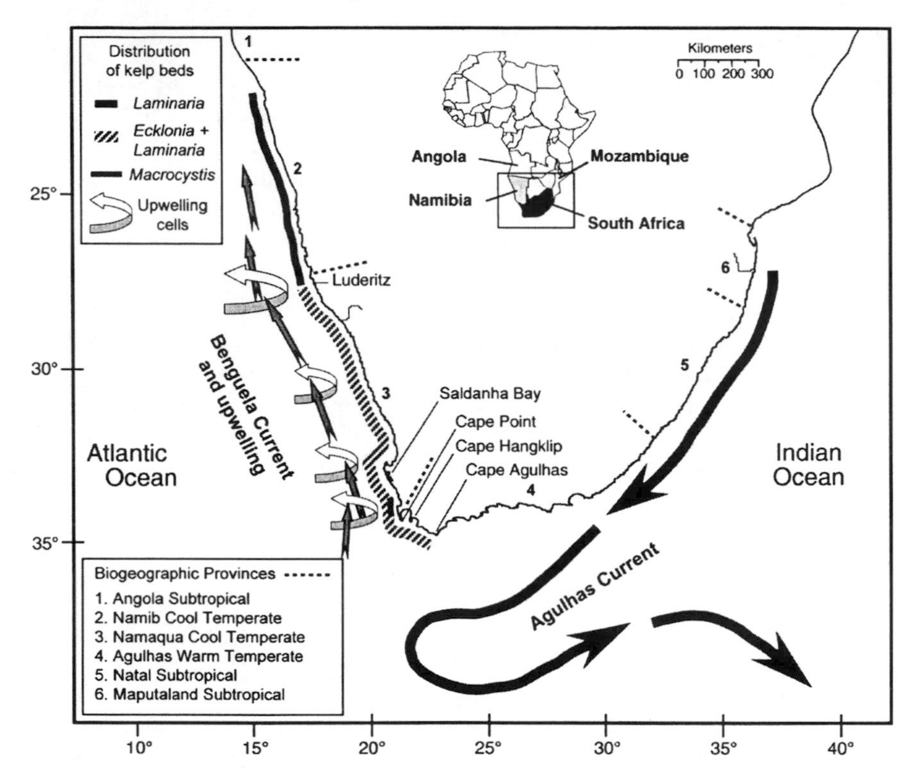

FIGURE 3.1 Map of southern Africa, showing localities mentioned in the text, the six biogeographic provinces and the distribution of the three species that form kelp beds (*Ecklonia maxima, Laminaria pallida,* and *Macrocystis angustifolia*).

Relationships to Physical Conditions

Wind plays a key role in the dynamics of kelp beds in southern Africa. In the Namaqua Province wind forces are strong and peak in summer (fig. 3.2a; Field and Griffiths 1991). Wave action is strenuous along most of the coast, reflecting the sheer coastline and long wind-fetch, with storms being generated up to 5000 km away in the South Atlantic. Waves exceed 6 m 10% of the time, and extreme waves exceed 18 m (Russouw and Russouw 1999). More importantly, wind direction changes seasonally, with offshore south-easterlies prevailing in summer and onshore north-westerlies in winter. As a result, upwelling is largely confined to summer, so that sea temperatures peak in winter and are practically a mirror image of air temperatures (fig. 3.2b). Wind is intermittent over most of the Namaqua Province, arriving in cycles that last about 7 days. Clear water is brought to the surface for periods normally lasting 3–4 days, introducing nutrients but exporting material from the kelp bed as the water moves offshore. After a lull, the wind reverses and offshore waters are

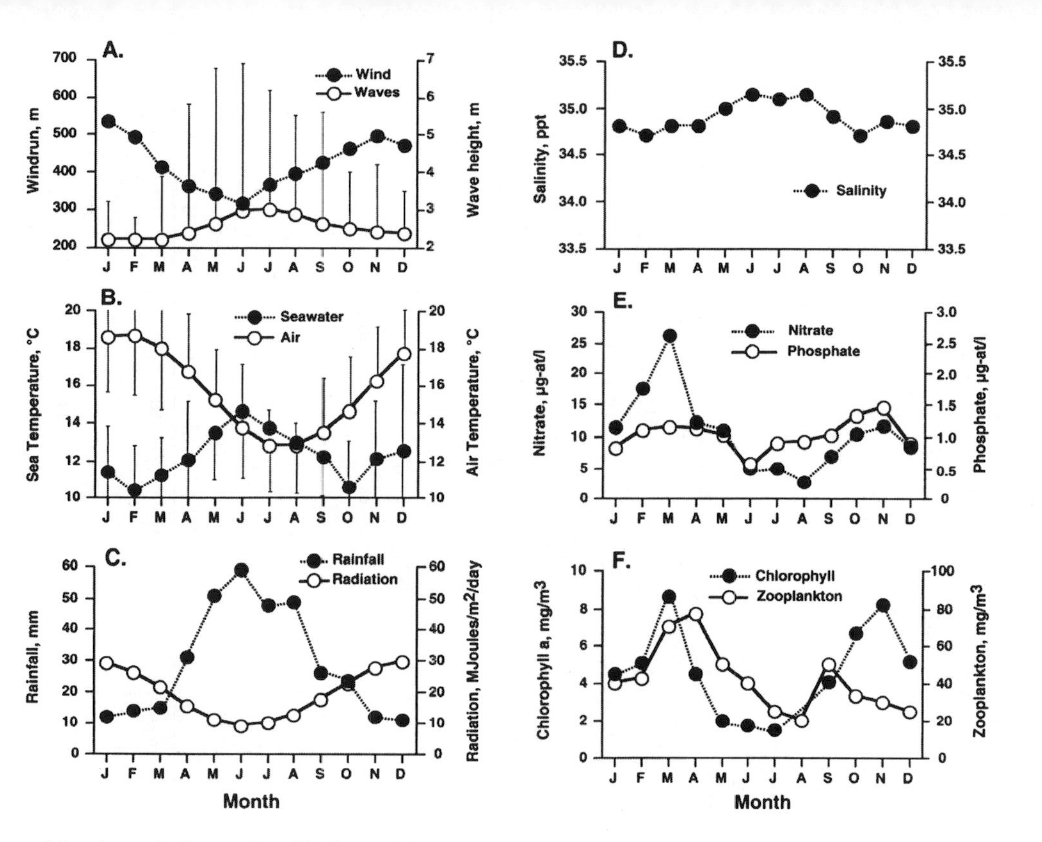

FIGURE 3.2 Monthly means of the physical, chemical, and biological characteristics of southern-Benguela kelp beds. Data derived from the two inshore stations listed by Andrews and Hutchings (1980), from Russouw and Russouw (1999), and from the South African Weather Service. Bars in (a) indicate mean maximum monthly values for wave height. Bars in (b) are standard deviations (some omitted or truncated to avoid overlaps). Data are provided in the units in which measurements were made. For phytoplankton, the ratio of total carbon to chlorophyll a averages 141 in winter and 65 in summer; and the ratio of grams carbon to kJ is 48 (Andrews and Hutchings 1980, Newell et al. 1982).

returned to the coast, introducing particulate organic matter (POM) and phytoplankton to the kelp bed, radically reducing light penetration. Day-to-day contrasts in conditions are the norm, with frequent temperature swings of 5°C–7°C. Topographic features concentrate upwelling in a series of six cells. At Lüderitz, upwelling is persistent, creating one of the largest and most intense upwelling cells in the world. This probably creates an environmental barrier responsible for the biogeographic differences between the northern and southern Benguela (Shannon 1985).

Tidal range is 1.8 m. Rainfall peaks in winter (fig. 3.2c) and is highest in the south, progressively declining to as little as 15 mm/yr in the Namibian desert because the cold upwelled waters yield little moisture. Salinity changes little, but does decline in spring and summer with intrusions of deep water (fig. 3.2d). Nutrient levels are high, peaking during summer upwelling (fig. 3.2e). Phytoplankton levels are extremely high and follow seasonal cycles of nutrient input (fig. 3f). One consequence of the high productivity is that ensuing decay often depletes oxygen, particularly in the north. Zooplankton is abundant (fig. 3.2f), but not exceptionally so, probably because the lifecycles of most species are not short enough to keep pace with the intermittent nutrient pulsing and phytoplankton blooms. Clearly, the responses of the kelp-bed community are powerfully influenced by upwelling and wave action.

The Food Web

Food Web Components

Primary producers Four sources of primary production fuel the food web: kelps, epiphytes, understorey algae, and phytoplankton. *Ecklonia maxima* and *Laminaria pallida* are the two most important contributors among the macrophytes, with production reaching 22,800 kJ/m^2/yr (fig. 3.3). The fronds of kelps are fast growing, achieving a production/biomass ratio (P/B) of 4.0. Growth peaks in summer, when both insolation and nutrient levels are high (Dieckmann 1980). Collectively, macrophytes produce 36,800 kJ/m^2/yr (Newell et al. 1982).

Phytoplankton has a relatively tiny biomass but, because of its high P/B, generates flows of 22,500 kJ/m^2/yr, not dissimilar to the production of macrophytes (Carter 1982, Newell et al. 1982). Immediately beneath the kelp canopy, phytoplankton production is reduced by 95% due to shading. Overall, this cuts 12% off the potential production of phytoplankton in the kelp bed (Borchers and Field 1981).

Total primary production of 62,200 kJ/m^2/yr in the kelp beds is equivalent to an energy conversion of 1.7% from incident

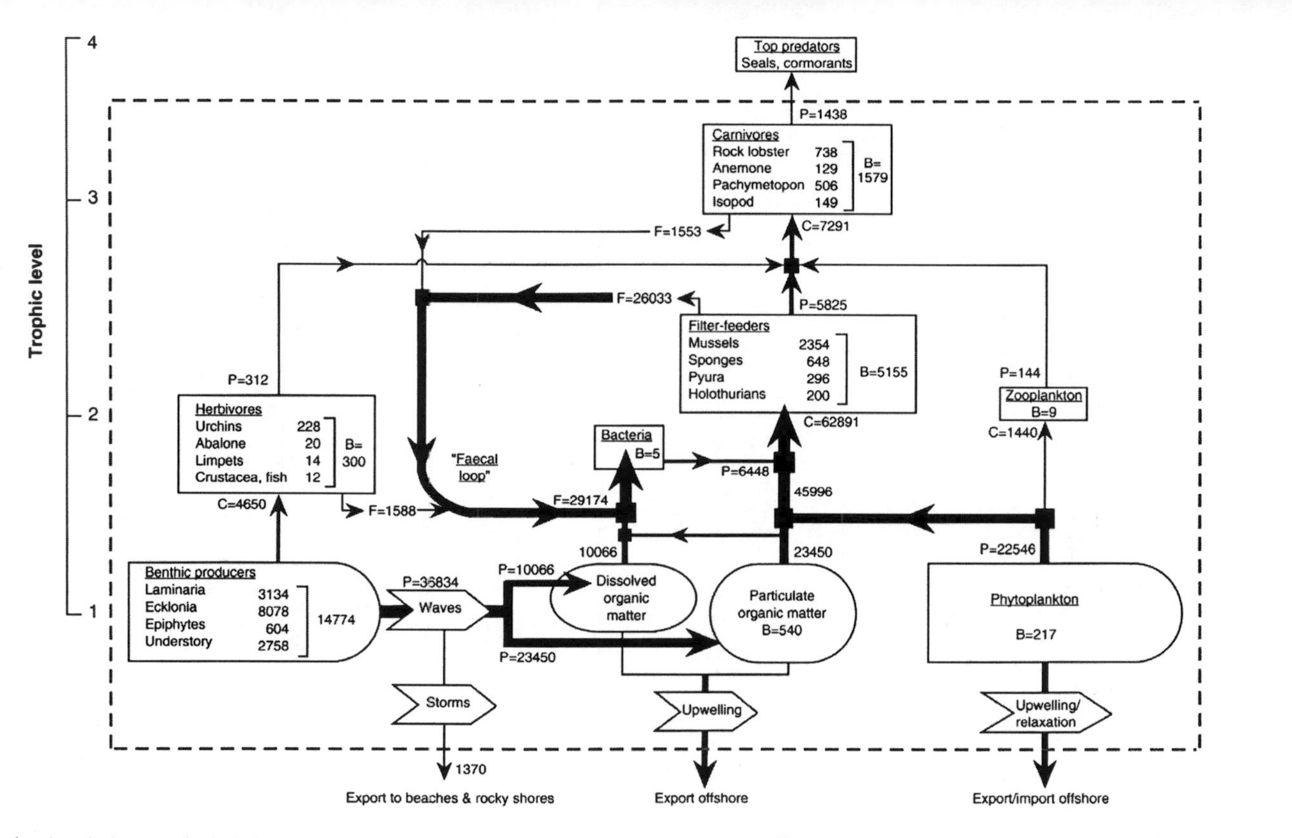

FIGURE 3.3 A food-web diagram for kelp beds on the west coast of the Cape Peninsula. Units of biomass (B) inside the boxes are in kJ/m^2. Flow rates on the arrows reflect production (P), consumption (C), or feces (F) in kJ/m^2/yr. Approximate trophic levels appear on the left. Widths of black arrows reflect the relative magnitude of flows. The dotted line indicates the kelp-bed boundary. Open arrows are "gateways" that influence processes. Data derived from Newell and colleagues (1982), Branch and Griffiths (1988), Bustamante and colleagues (1995b), Bustamante and Branch (1996). For approximate conversions, ratios of wet:dry mass, dry mass:carbon, and grams carbon:kJ are 0.15, 0.29, and 54 for benthic algae, and 0.20, 0.35, and 65 for invertebrates. Detailed conversions for individual taxa appear in Field and colleagues 1980.

illumination—a figure that Newell and colleagues (1982) suggest is close to the maximum attainable by aquatic plants. This total can be compared to a value of 54,000 kJ/m²/yr for phytoplankton alone in open waters nearby the kelp beds. There is close agreement between the primary production in the kelp bed and the rate of primary consumption (67,500 kJ/m²/yr), implying that that the two are in equilibrium or that any exports from the kelp beds are balanced by imports.

Three key "gateway" processes influence the fate of primary producers. First, waves continually erode kelps and other macrophytes, 62% of their production being directed into particulate organic matter (POM) and 26% into dissolved organic matter (DOM; Newell et al. 1980). Second, upwelling and downwelling play key roles in the export or import of this material and of phytoplankton. Third, storms detach and distribute whole plants, thus removing about 6% of the production, most of which finds its way to sandy beaches and rocky shores. Such subsidization is of central importance in these ecosystems, contributing large proportions of the energy and material flows (Koop and Griffiths 1982, Branch and Griffiths 1988). Intertidal rocky shores on the west coast support the highest biomasses of grazers recorded anywhere in the world, and their maintenance depends on the importation of kelp debris and access to kelp blades (Bustamante et al. 1995b). Intertidal filter feeders also derive the majority of their food from particulate kelp (Bustamante and Branch 1996).

Primary consumers Within the kelp beds, the majority of the POM and the production by phytoplankton are consumed by filter feeders, which account for 77%–89% of the total animal biomass and productivity. By far the most important filter feeder is the ribbed mussel *Aulacomya ater*, but other notable contributors are sponges, the solitary ascidian *Pyura stolonifera*, holothurians, and the black mussel *Choromytilus meridionalis*. *A. ater* is sufficiently abundant to completely filter a column of water 10-m deep in 7.5 hours. Mussels are capable of filtering and absorbing kelp particles, phytoplankton, bacteria, and faecal matter with absorption efficiencies of 50%–69% (Stuart et al. 1982a,b; Seiderer and Newell 1985), although samples in the vicinity of mussel beds suggest that detrital particles make up 85% of the food. The age of detrital particles affects the scope-for-growth by mussels, yielding greatest returns at an age of 2.8 days—which coincides with the average age of particles in the kelp bed (Stuart 1982).

Zooplankton is a relatively trivial player in kelp beds, although offshore it assumes a dominant role, reaching a biomass of 8%–15% of that of phytoplankton. There are two reasons why zooplankton is seldom important in the kelp bed. First, upwelling exports the zooplankton faster than the lifecycles of constituent species allow populations to build up. Second, even offshore where zooplankton does have the opportunity to build up its populations, its slow turnover, coupled with the intermittent nature of upwelling, often means that the zooplankton does not have time to capitalize on phytoplankton blooms before upwelling is reversed and blooms decline. This "match/mismatch" situation means that zooplankton levels are seldom as high as might be expected from the high primary production (Andrews and Hutchings 1980).

One of the surprises about the kelp-bed food web is that herbivores account for very little of the benthic primary production—perhaps 12% (Newell et al. 1982). Moreover, the two most important herbivores, the sea urchin *Parechinus angulosus* and the abalone *Haliotis midae*, derive most of their food by trapping fragments of seaweeds, so they are not grazers (Barkai and Griffiths 1986, Day and Branch 2002a). The consequences of this are explored below. Other herbivores that are true grazers include the fish *Sarpa salpa*, winkles (*Turbo* and *Oxystele* spp.), limpets, amphipods, and isopods. None of these ever contributes substantially to the animal biomass or production, but they do check the growth of benthic macroalgae by feeding on sporelings and therefore promote encrusting corallines.

Predators There are five main guilds of predators in the kelp beds, each with distinctive effects. Anemones, which are sit-and-wait predators, constitute 8% of predator biomass and rely on water-transported prey to come within reach and thus have little effect on prey populations. There is a striking scarcity of predatory fish and the most abundant fish, *Pachymetopon blochii*, which contributes 32% of predator biomass, feeds omnivorously on a large variety of invertebrates combed from epiphytic algae (Branch and Griffiths 1988).

The commercially harvested rock lobster *Jasus lalandii* is the dominant predator, contributing 41% of the predator biomass. Its main prey is the ribbed mussel *A. ater*, but urchins, polychaetes, crustaceans (especially other rock lobsters), sponges, and even algae also feature in their diet. Barkai and Branch (1988a) showed that rock lobsters consume barnacles shortly after their settlement, and barnacles constitute a major source of food, even although their standing stocks are scarcely detectable. In areas where rock lobsters are exceptionally abundant, their energetic needs may reach 36,000 kJ/m^2/yr and they eliminate virtually all macro-prey. Barnacle recruits, which can produce up to 25,000 kJ/m^2/yr, then become an essential part of their diet. This emphasizes the vital role that small and often overlooked species play in maintaining energy flows because of their high P/B ratios.

J. lalandii has strong effects on prey abundance and size composition. In particular, it seems responsible for a bimodal size structure in populations of the mussel *A. ater*. One mode comprises individuals <25 mm in length and the other individuals 50–90 mm; there are practically no mussels of 25–50 mm. More than 85% of the mussel biomass comprises individuals over 60 mm. This has profound effects on rock lobsters, because if they are smaller than 80-mm carapace length they are incapable of feeding on mussels that exceed 60 mm (Pollock 1979) and are at a competitive disadvantage relative to larger rock lobsters (Griffiths and Seiderer 1980). This reduces rock-lobster productivity and yields. Seiderer and colleagues (1982) modelled the potential consequences of harvesting small lobsters to thin their densities and concluded that this would increase mussel biomass by 87%, improve rock-lobster yields 19%, but might reduce their egg output by 35%. There is no easy way to forecast the long-term consequences of the latter.

The fourth guild of predators constitutes small predators, with the isopod *Cirolana imposita* serving as an exemplar. Biomass of this small predator is low but, because of its high turnover (P/B = 5.6), it contributes more to energy flow than all the other predators combined: 2230 kJ/m^2/yr (Shafir and Field 1980). Lastly, there are nonresident predators that transiently visit and feed in kelp beds. Seals (*Arctocephalus pusillus*) and bank cormorants (*Pholacrocorax neglectus*) are notable examples, preying on rock lobsters and small fish.

The role of bacteria The microscopic size of microheterotrophs and the specialized techniques needed to work on them have hampered understanding of their role in food webs, but studies on their function in kelp beds have done much to lift this veil of ignorance. Bacterial numbers in kelp beds are extremely high. Just the erosion of kelp fronds contributes 2.6×10^9 bacteria to each square meter of water column per day. Bacterial biomass is minute (5 kJ/m^2), but because of their exceptional productivity (P/B ≈ 1000), they generate about 6450 kJ/m^2/yr—close to 10% of the needs of filter feeders (Newell et al. 1982, Newell and Field 1983).

The first role played by bacteria is as a link between DOM released by algae, and filter feeders. Nearly all this DOM is taken up by bacteria, and converted into bacterial biomass with an efficiency of 6%–33%, depending on the particular components involved. Mannitol, the major photosynthetic product of kelps, is metabolized extremely fast by bacteria. Koop and colleagues (1982) used the presence of mannitol-fermenting bacteria to track the export of materials from kelp beds, which they estimate can be transported up to 9 km offshore.

Bacteria also colonize and metabolize POM released by algae (Stuart et al. 1981), but this material is also directly consumed by filter feeders. Bacteria are themselves ingested and digested by mussels, which produce a bacteriolytic enzyme that is 10,000-fold more active when temperatures drop to 10.0°C–11.5°C. This coincides with conditions experienced during upwelling, when most sources of food diminish because of export. Consequently, digestion of bacteria may be vital at such times (Seiderer and Robb 1986). Bacteria rapidly colonize and metabolize feces (Stuart et al. 1982b). Because of the losses of carbon that take place in the process, bacteria cannot contribute more than about 10% of the carbon and energy needs of filter feeders; but they reduce the C:N ratio of feces down to 2.7:1, thus creating a concentrated source of nitrogen. By comparison, detritus has a C:N ratio of about 18.9:1, and phytoplankton a ratio of 6.6:1. Because of this, bacteria potentially provide 73% of the nitrogen needs of consumers in kelp beds. Moreover, if filter feeders consume nitrogen-enriched feces, they will in turn produce more feces that can be colonized by bacteria and then recycled. Newell and Field (1983) call this a "fecal loop," which collectively transfers about 29,200 kJ/m^2/yr to bacteria (fig. 3.3). With a conversion efficiency of 15%, this bacterial production contributes 43,700 kJ/m^2/yr to the energetic needs of filter feeders.

The final important role of bacteria is their rapid remineralization of algal fragments and feces, which contributes to the nutrient pool. During upwelling, nutrients may be present in superfluity; but during downwelling local remineralization assumes greater importance.

Food Web Processes

Kelp beds as open systems The model of kelp-bed dynamics assumes a closed system—or at least one in which imports are in equilibrium with exports (fig. 3.3). The close correspondence between annual production and consumption creates confidence in the budget, but the fact remains that no account is taken of short-term fluxes associated with the pulsed upwelling characteristic of the southern Benguela. To explore this effect, Wulff and Field (1983) simulated upwelling and downwelling to determine their effects on energy flows to filter feeders (fig. 3.4). Four patterns emerged. First, during downwelling, phytoplankton is imported and makes up 40%–80% of the food consumed, whereas during upwelling its contribution is negligible and kelp POM and feces dominate. Second, the total amount of food available during upwelling is never enough to sustain the needs of the filter feeders, but during downwelling there is excess food. Third, the relative contributions of different food sources are strongly affected by the rate of water turnover. Fourth, the proportion of filter feeder requirements met decreases as the frequency of upwelling rises, but increases with the frequency of downwelling.

Pulses of upwelling are common in southern Benguela, kelp beds. During summer, the southeasterly winds blow more frequently and intensely, and there are cycles of about 3–4 days of upwelling, followed

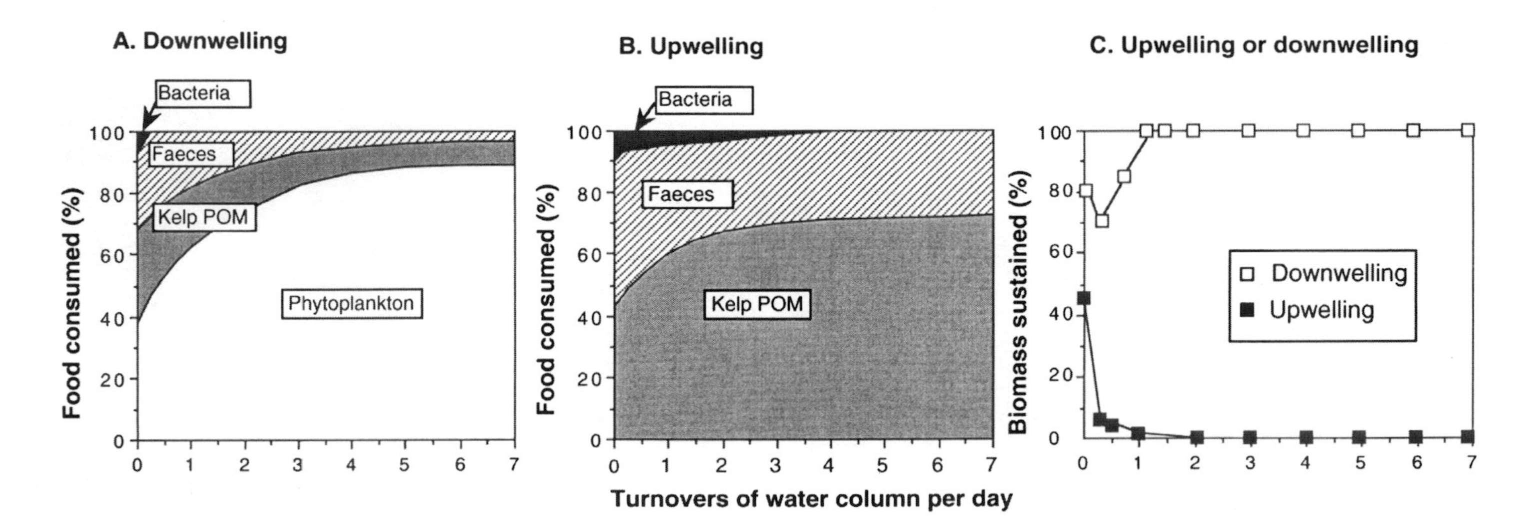

FIGURE 3.4 Output of models predicting the relative importance of bacteria, feces, particulate kelp, and phytoplankton in the diet of kelp-bed filter feeders under conditions of (a) downwelling and (b) upwelling, in relation to the rate of water exchange (turnovers per day). Panel (c) summarizes the proportions of filter-feeder standing stocks that can be sustained during downwelling or upwelling. Modified from Wulff and Field (1983).

by 3–4 days of relaxation and downwelling. In winter north-westerlies prevail and upwelling occurs only 10% of the time. Peak productivity by both phytoplankton and kelps occurs in summer, when light intensities and nutrient levels are high. Ironically, it is only during winter that filter feeders experience a positive scope for growth, because export of materials by upwelling deprives them of food in summer.

Nontrophic effects on community dynamics Analyses of energy and material flows through ecosystems are good for distilling trophic interactions among species, but fail to capture a range of nontrophic processes (fig. 3.5). For example, kelps lie at the base of the food web, but they also fulfill functions other than primary production. One example is that kelp holdfasts serve as important nurseries for algal sporelings, particularly those of kelps themselves (Anderson et al. 1997). Disproportionately large numbers of sporelings are associated with holdfasts, and the ratio of sporelings on rocks to those on holdfasts is inversely correlated with the density of grazers, implying that holdfasts provide better protection against grazing. This has important implications for harvesting. Holdfasts soon decay and disappear if kelp plants are cut at the stipe. Harvesting in this manner removes thus the plant and the holdfast upon which future recruitment is dependent.

Another example is that water movements sweep the blades of *Laminaria pallida* across the rock face, creating a grazer-free swathe around clumps of plants, within which young sporophytes can become established (Velimirov and Griffiths 1979). *Ecklonia maxima* has floats that hold its fronds at the surface, so it does not have the same sweeping effect.

Kelps contain polyphenols, which have long been held to deter herbivores because they inhibit digestion by binding with digestive enzymes. In line with plant-defense theory, Tugwell and Branch (1989) hypothesized that polyphenols should be invested mainly in tissues that are particularly important for survival, and should also be concentrated in the outer layers of tissues. Indeed, polyphenols have proven to be more concentrated in meristems, holdfasts, stipes, and sporogenous tissues than elsewhere, and there are exceptional concentrations of up to 40% dry mass in the outer meristoderm layers, which will be the first point of contact when a grazer attacks.

Although polyphenols were initially widely considered as a means of repelling herbivores, the idea was called into question when counter-adaptations to the precipitating effects of polyphenols were discovered, including gut surfactants (see Tugwell and Branch 1992 for references). The apparently ubiquitous presence of surfactants in the guts of animals was a particular challenge to the anti-herbivore role of polyphenols. Tugwell and Branch (1992) explored the efficacy of surfactants in several species, and showed that in some they do indeed nullify polyphenols, but in others they have no effect. In

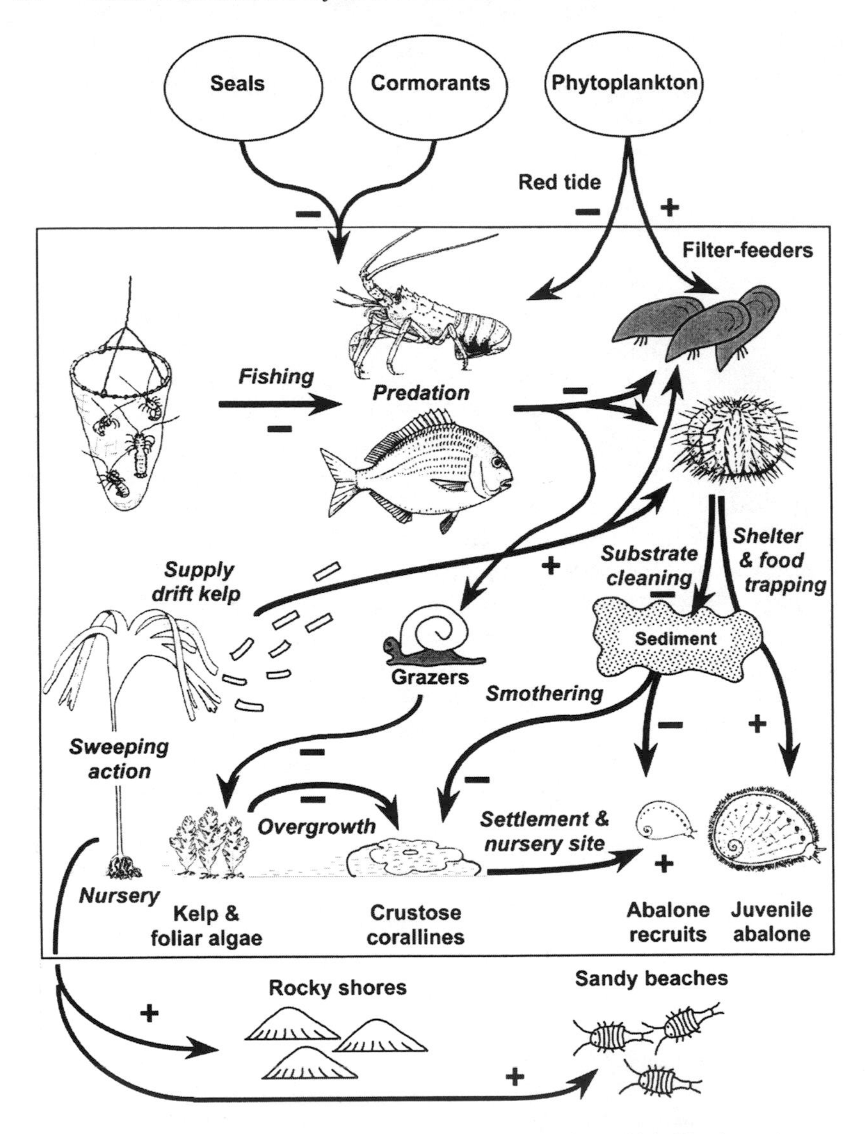

FIGURE 3.5 Interactions in southern Benguela kelp beds. Italicized labels indicate the nature of the process; arrows point toward the component affected by the interaction; + or – signs indicate whether the interaction has a positive or negative effect. Modified from Day and Branch (2002a).

particular, the isopod *Paridotea reticulata*, which feeds preferentially on the sporogenous tissues of kelps, is immune to the effects of polyphenols. Thus, the effects of surfactants are species-specific: only in some cases do they neutralize polyphenols. Because kelp POM is consumed by filter feeders, it would be useful to know whether polyphenols retain their efficacy after fragmentation and decay.

Although they are not harvested, the role of sea urchins in kelp beds has become of central interest. In many parts of the world, they are dominant grazers and control the growth of macroalgae. In the southern Benguela, however, urchins fulfil a different role. Large-scale field experiments involving removals of the urchin *Parechinus angulosus* failed to yield any changes in the composition or abundance of macroalgae, including kelp sporelings (Day and Branch 2002a). The apparent reason was that *P. angulosus* is normally not a grazer but a trapper of algal debris—particularly kelp debris. Removal of urchins did, however, have several other effects, some completely unanticipated. Previous work showed that almost all juveniles of the abalone *Haliotis midae* (90%–98%) hide beneath urchins, gaining protection from predators (Tarr et al. 1996). This led to an expectation, subsequently verified, that removal of urchins would radically reduce the abundance of juvenile abalone (Day and Branch 2002a). But there were also several unexpected effects of urchin removal. First, freshly settled "recruits" of abalone (< 2 mm) are not obviously associated with urchins (Day and Branch 2000b), yet they too virtually disappeared after urchin removal, possibly because sedimentation increased after urchin removal—in itself an unexpected outcome. Another surprise was that kelp debris disappeared from the floor bed once urchins were eliminated (Day and Branch 2002a). Evidently, urchins play a central role in maintaining kelp debris in the system by trapping fragments. In the process, juvenile abalone concealed beneath urchins gained not only protection but also access to a free meal in the form of this trapped kelp debris. Juvenile abalone held in aquaria with urchins alone take up half as much food as those held with urchins and kelp, and are forced to leave the shelter of urchins to forage for alternative food (Day and Branch 2002b). Urchins thus serve as mobile, protective cafeterias for juvenile abalone.

Although *P. angulosus* seems incapable of controlling macroalgae on its own, experimental exclusion of all benthic herbivores (including *P. angulosus*) does result in a proliferation of frondose algae and a consequent overgrowth of encrusting corallines. Abalone larvae settle selectively on encrusting corallines (Day and Branch 2000b) and their settlement is likely to decrease in the absence of urchins. For the lucrative abalone-harvesting industry, the complex web of interactions in kelp beds thus has a profound significance.

Temporal and Spatial Variations

Depth-Related Differences

Surveys of kelp-bed communities show two patterns. The first reflects a stratification of communities by depth. *E. maxima* is the dominant kelp in inshore waters of 2–8 m depth where it contributes 50%–70% of the kelp biomass. As depth increases, *L. pallida* becomes more important, contributing 40%–70% of biomass at intermediate depths of 8–9 m. Closer to shore it forms a subcanopy beneath *E. maxima* but it often becomes the sole kelp in deeper water. Offshore at 10–20 m kelps are absent or patchy but animals become much more diverse and dominate the biomass (Velimirov et al. 1977, Field et al. 1980). Two overriding factors dictate these depth-related patterns: diminishing light and reduction of wave action.

Geographic Patterns

The second spatial pattern concerns differences between regions. *L. pallida* is the only kelp present in the Namib Province, whereas both *E. maxima* and *L. pallida* occur in the Namaqua Province. In the southern Benguela, kelp biomass increases from north to south, but mussel and rock lobster biomasses decrease (Field et al. 1980). This reflects increasing water turbidity in the north, which cuts down light penetration and thus reduces the depth to which kelps penetrate, but increases the amount of food in suspension for filter feeders.

Rock Lobsters: Causes of Alternative Stable States?

Comparisons of the composition and functioning subtidal reefs at two adjacent islands, Marcus and Malgas Islands, which lie in the mouth of Saldanha Bay (see fig. 3.1), have raised particular challenges concerning the top-down effects of predators. Specifically, these two systems seem to be alternate stable states that are maintained by biotic interactions, particularly the predatory effects of the rock lobster *Jasus lalandii* (Branch et al. 1987, Castilla et al. 1994). Malgas Island supports a dense population of rock lobsters (10/m^2, with a wet biomass of 3.9 kg/m^2) and kelps and other benthic algae are abundant. The black mussel *Choromytilus meridionalis* is, however, absent and the ribbed mussel *Aulacomya ater* scarce. Animal diversity as a whole is low. In contrast, Marcus Island has virtually no rock lobsters or benthic algae, but thick beds of *C. meridionalis* and a diverse array of other animals. Castilla and colleagues (1984) argue that these differences are maintained by the presence or absence of dense rock-lobster populations. Experimental exclusion of rock lobsters by cages at Malgas led to the development of communities dominated by black mussels, similar to those at Marcus. Moreover, black mussels that colonized cages at Malgas survived as well and grew as fast as those at Marcus Island. This implies that physical conditions and food supplies were comparable at the two islands and unlikely to explain the radical differences in community composition (Barkai and Branch 1988b, 1989). Thus, intense predation by rock lobsters at Malgas and its absence at Marcus remain the parsimonious explanation for the observed differences.

Few animal species survive at Malgas Island, except for those that have defences against the rock lobsters. These include hydroids that have stinging cells, the whelk *Argobuccinum pustulosum*, which has a thick shell, and large individuals of the ribbed mussel *A. ater*. But the most interesting survivor is the whelk *Burnupena papyracea*, which is coated by the bryozoan *Alcyonidium nodosum*, which deters rock lobsters (Barkai and McQuaid 1988).

A massive transplant of 2000 rock lobsters from Malgas to Marcus resulted in the most remarkable and unexpected outcome. Dense populations of the whelks *Burnupena* spp. (200/m^2) overcame and consumed the transplanted rock lobsters within a matter of minutes. In a second experiment, smaller numbers of rock lobsters were again transferred to Marcus, but held in cages that excluded the whelks. Thus protected, the rock lobsters survived but once released from their cages they suffered instant consumption by the whelks. This is an extraordinary reversal of predator–prey relationships, because rock lobsters normally consume *Burnupena*, and at Malgas Island, only those *B. papyracea* that manage to grow large enough to acquire a protective covering of *A. nodosum* can survive—and their densities never exceed 0.5/m^2.

Is this a case of alternative stable states? Connell and Sousa (1983) are skeptical of the concept, arguing that three criteria are necessary for its acceptance: (1) sufficient time for the "stability" to exceed the lifespan of all individuals; (2) comparability of physical conditions; and (3) natural self-perpetuation and an absence of human intervention. Castilla and colleagues (1994) believe that all three criteria have been met in this situation because differences have persisted for at least 22 years and both localities are in a marine protected area with no easily discernable differences in physical conditions. This does seem to be one of the few valid cases of alternative stable states, even by the stringent criteria proposed by Connell and Sousa. At the same time, the differences between the islands should not be overgeneralized. The contrasts are striking and do seem explicable by the biotic processes that have been unravelled; but they also seem localized—pairings of such extremes of community composition have not been recorded elsewhere.

Biological Regime Shifts?

Studies on the dynamics of rock lobsters have also led in other directions. Long-term measurements have shown that there has been a substantial reduction in the growth rate of *J. lalandii*, profoundly reducing its productivity. Growth has always been slower in the north, with females maturing at 57–59 mm carapace length, as opposed to 66 mm in the south. Newman and Pollock (1974) showed that growth was slow in areas where the benthic faunal biomass, particularly that of the mussel *A. ater*, was low. Since the 1980s, however, growth rates have declined more generally, dropping from about 4–8 mm/yr to 1–2 mm/yr (Pollock 1986, Pollock and Shannon 1987). Three possible explanations are that (1) as fishing is directed at the larger individuals in the population, there may have been evolutionary selection for slow growth and early maturation; (2) food may have diminished; and (3) changes may have taken place in physical conditions, linked to environmental change. Current evidence is circumstantial and suggestive but equivocal. Mayfield and colleagues (2000) showed that gut fullness and dietary composition did not differ appreciably among fast-growth and slow-growth areas, despite clear differences in the availability of food. They argued that where food is short, rock lobsters will spend more time and effort foraging, thus reducing their growth even though they succeed in filling their guts.

An alternative hypothesis is that there has been a progressive increase in the extent of oxygen-deficient waters on the shelf of the west coast, forcing rock lobsters to concentrate inshore where competition due to crowding depresses growth (Pollock and Shannon 1987). Two processes are advanced as root causes: (1) upwelling may have increased, enhancing nutrient inputs and boosting phytoplankton production and (2) removal of large amounts of planktivorous pelagic fish by fishing may have diminished grazing on phytoplankton. Separately or together, these changes may have increased decay and diminished oxygen.

A second change in rock-lobster populations has been an extension of their range southeast of Cape Hangklip. In the late 1970s rock lobsters were rare there, but by 1997 they had achieved substantial standing stocks of 652 kJ/m^2 (about 143 g wet-mass/m^2) and large average sizes. Only rock lobsters exceeding 68-mm carapace length are capable of eating urchins, and a negative correlation exists between the densities of such large rock lobsters and urchins (Field et al. 1980, Mayfield and Branch 2000, Mayfield et al. 2001).

The increase in rock lobster densities has been associated with substantial changes in the community, including a 95% depletion of herbivores such as urchins and the winkle *Turbo cidaris*, and an 800-fold increase in foliar algae, including kelps. The decline in urchins has, in turn, been linked to the near-demise of juvenile abalone (Tarr et al. 1996), so that this major ecological shift constitutes a catastrophe-in-the-making for the abalone industry, the heart of which lies in this region. Farther to the east, where rock lobsters are still rare or are absent, the community structure remains unchanged, with abundant herbivores, small amounts of foliar algae and normal densities of abalone juveniles (Tarr et al. 1996).

To complicate the story, the stoking of phytoplankton blooms by upwelling is periodically followed by calm periods when the blooms become entrapped in bays and nutrients become limiting. Resultant decay of the blooms depletes oxygen, sometimes generating toxic hydrogen sulfide ("black tide") and causing mass mortalities of marine life (Matthews and Pitcher 1996). Rock lobsters mass inshore where turbulence provides some relief from the anoxia, but this often results in "walkouts," with up to 2000 tons of rock lobsters becoming stranded on the shore (Pitcher 1999, Cockcroft 2001). Afterwards, there are periods when urchins become so dominant that they occupy 100% of the reef floor. Although localized in two particular bays on the west coast of South Africa, such events completely transform the benthic community.

These effects are striking reminders that variations in winds powerfully affect kelp-bed communities. Ultimately, these bottom-up effects dictate patterns of productivity, whereas top-down effects determine community composition. Some of the resultant ripple effects were almost impossible to predict, and many of the twists and turns in the story could never have been elucidated without a combination of long-term monitoring, broad-scale surveys and detailed field experiments.

Human Influences

Mining, mariculture, alien species, pollution and global environmental change have all been identified as threats to intertidal and subtidal reefs (Branch et al. 2008). In the context of kelp beds in South Africa, however, the major threat is overexploitation of living resources.

Pollution and Climate Change

There are no point sources of pollution that affect kelp beds to any significant extent. Those that do exist, such as factory wastes, sewage outfalls, and organic effluent from fish factories, are situated in metropolitan Cape Town or in fishing villages and have only local effects. Oil spills are an ongoing threat, particularly to marine birds, but have had no recorded effect on subtidal reefs, other than short-term cessation of feeding by rock lobsters (Glassom et al. 1997).

Global climate changes may be influencing the dynamics of subtidal reefs, although the evidence is not conclusive. Several observations are, however, extremely suggestive. These are (1) the spread of rock lobsters at the southern end of their range (Mayfield and Branch 2000); (2) a parallel spread in kelp beds; (3) decreased growth rates of rock lobsters (Pollock 1986); (4) greater frequencies of lobster "walk-outs"; (5) possible intensification of upwelling (Pollock and Shannon 1987); (6) an apparent increase in the frequency of harmful algal blooms (Pitcher 1999); and (7) periodic "warm events" (Branch 1984).

Alien Species and Mariculture

The coast of South Africa is relatively free of alien invasive species. Two species constitute a threat to intertidal systems: the European green crab *Carcinus maenas* and the Mediterranean mussel *Mytilus galloprovincialis*; but neither has any effect on subtidal reefs (Griffiths 2000a, Branch and Steffani 2004).

Mariculture is being actively pursued in South Africa, but is confined to calm-water bays on the west coast (mussel culture and growth of the agar-producing alga *Gracilaria gracilis*) or land-based operations (for the abalone *Haliotis midae*). The greatest threats from mariculture lies either in the escape of introduced species that have the potential to become invasive, or the introduction of associated diseases and pests. Neither effect has manifested itself on any subtidal reefs in South Africa, although the introduction of *M. galloprovincialis* on the south coast did result in its escape and colonization of intertidal shores (Branch and Steffani 2004). Attempts to rear the South African abalone *H. midae* in California led to the introduction and escape of a sabellid pest that drills into shells and reduces the growth rate of hosts. This led to the closure of most abalone farms there, and strenuous efforts were required to eradicate the pest (Kuris and Culver 1999). Dangers do therefore exist, but so far subtidal reefs in South Africa remain unscathed.

Mining

Mining for diamonds has unique effects on the west coast of southern Africa. Two different types of operations are involved. First, land-based mining produces large amounts of "fines" or discarded sediments. In the southern part of Namibia there are places where these fines are disposed of on beaches. Apart from causing smothering and beach accretion, these sediments become suspended and cause local high-turbidity plumes, and their deposition on sheltered reefs increases densities of filter feeders and diminishes grazers, promoting opportunistic algae (Pulfrich et al. 2003b).

Second, divers operating subtidally pump up gravel with long suction pipes that deposit the gravel on the shore where it is sorted. This has a range of effects. Kelp is cut to allow the divers access. Removal of gravel destabilizes boulders. Deposition of waste material scours the shore. The net effect in the subtidal zone is a reduction of species richness and abundance, particularly of foliar algae and filter feeders (Pulfrich et al. 2003a).

In both types of operation, the effects are localized and recovery takes place within two years. Specific concerns have been raised about the effects of diamond mining on rock lobsters, because of potential clashes of interest between mining and fishing. However, surveys comparing the abundance, size, and sex ratios of lobsters in mined and control areas failed to detect any significant effects.

Harvesting

Prehistoric harvesting of intertidal animals extends back farther in Africa than anywhere else in the world and includes Eritrean middens dating to 125,000 years ago (Walter et al. 2000). Both the abalone *Haliotis midae* and the rock lobster *Jasus lalandii* feature in South African middens from about 10,000 years ago onward (Jerardino and Navarro 2002; T. Peschak, pers. comm.). One of the more striking features of the rock lobster remains is that the modal size of harvested specimens was 120–140 mm carapace length (A. Jerardino, pers. comm.), close to double that in modern commercial harvests and an indication of the probable effects of recent harvesting on size structure.

Kelp has been commercially harvested for the past 40 years, first in the form of beach-stranded plants used to extract alginate, and then as live plants used to obtain agricultural growth stimulants or for abalone food in mariculture. Removal of beach-cast material does compromise the ecology of intertidal sandy beaches because it reduces beached debris. Kelp beds harvested by cutting plants do, however, recover within two years (Levitt et al. 2002).

The stocks that have been most affected by harvesting are the rock lobster *J. lalandii*, the abalone *H. midae*, and a number of linefish. Rock lobster harvests have declined from a peak of 12,500 down to 2,800 tons/yr as the resource has been mined out by commercial fishing. Part of this decline was inevitable, given the slow growth of the species. Harvests stabilized at about 4000 tons/yr during the 1980s before sustained reductions in growth rate took their toll, necessitating reductions in total allowable catches (Pollock 1986, Crawford et al. 1987).

Rock lobsters are still relatively well managed, but the same cannot be said for abalone, which are in a catastrophic condition due to the dual effect of increased predation by rock lobsters and intensive poaching (Hauck and Sweijd 1999). Over the past century, many species of linefish have been fished down to 5%–10% of pristine levels (Griffiths 2000b). In general, resources that are both lucrative and occur near to the shore have proven the most difficult to manage, as greed and accessibility conspire to make them vulnerable.

Management Measures

Four different (and often overlapping) styles of management have been applied to nearshore and intertidal resources in South Africa. First, some species have benefited from detailed single-species studies and management procedures. Two examples demonstrate both the advantages and the pitfalls of this approach. When kelps are harvested, recovery is quick because sporelings settle and grow rapidly in areas where canopies have been removed. There also appear to be no effects on benthic understory species. Levitt and colleagues (2002) have, however, shown that the recovery of kelp is greatly influenced by the mode and frequency of harvesting. Regrowth, and therefore yields, are maximized if harvesting is repeated four-monthly and if the kelp fronds are cut at distances of 20–30 cm from their bases (as opposed to removing whole plants). This procedure has the added advantage of leaving plants intact, retaining the nursery function of holdfasts. Ecological effects can thus be minimized and returns maximized.

A second successful case has been the application of "Operational Management Procedures" (OMPs) to the rock lobster *Jasus lalandii* (Johnston and Butterworth 2005). OMPs involve the development of an algorithm that is agreed upon in advance by all parties, including industry, managers, and scientists, and can be used to determine the total allowable catch (TAC) in any given year. In relation to rock lobsters, three important variables are used in the algorithm: growth rate in the prior year; catch-per-unit-effort recorded from commercial fishing, and a measure of abundance obtained from fisheries-independent monitoring. Models can be developed to predict the consequences of a range of TACs. Objectives can be set in advance (such as allowing for a recovery in the stock), and a TAC chosen to meet the objectives. The advantages of this procedure are that it allows mutual prior agreement about objectives and the parameters used in the OMP, and it establishes a relatively objective means of making decisions. The proof of the pudding is that rock lobster stocks have built up, and TACs consequently increased, since implementation of the OMP.

Such approaches are, however, inherently limited because the real world of kelp beds involves a plethora of interactions that are not easily factored into quantitative single-species modelling. In the kelp bed alone, there are four sets of exploited species, which include rock lobsters, abalone, kelps, and linefish, not to mention other species such as urchins that fulfill vital roles. Awareness of these complexities has promoted a second approach, based on ecosystem management (Cochrane et al. 2004). This is easier said than done, because multispecies models are not yet at the point where they can usefully predict optimal catches. Nevertheless, energy-flow studies and field experiments on species interactions in kelp beds have expanded our understanding and compelled a consideration of the ecosystem as a whole. Nowhere is this more obvious than east of Cape Hangklip, where increases in the abundance of rock lobsters threaten the existence of abalone.

A third approach that is still in its infancy in South Africa has been comanagement: the joint involvement of users and managers in decision-making and management of resources. This approach finds its greatest application in near-shore resources that can be harvested by almost anyone using low-tech equipment. The problems of managing such resources are legion, and management is likely to fail if it does not have the approval of users. Recently, particular attention has been directed at the plight of subsistence fishers in South Africa (Harris et al. 2002), and how comanagement has been actively and successfully implemented for them on the east coast of South Africa (Harris et al. 2003). In theory it should be just as desirable to implement comanagement for industrial fishers, but in practice progress has been sluggish.

The fourth approach to management has been the creation of marine protected areas (MPAs). Properly implemented, MPAs have proven dramatically successful in restoring depleted populations (Buxton and Smale 1989, Attwood and Bennett 1995). The lessons learned have also found application in the development of criteria for evaluating the relative value of different areas as MPAs. These are based on a set of objectives that embrace conservation, fisheries management, and other human uses, and integrate environmental and social considerations. Stipulated criteria evaluate how effective areas are in meeting these objectives. At the heart of the system is the desire to obtain protection for representatives of all habitats in all biogeographic regions (Hockey and Branch 1997, Roberts et al. 2003a,b,).

Summary

Research on kelp beds in South Africa has gone though three phases. First, descriptive surveys revealed the high biomass, high productivity, and low diversity of the west coast compared to the east coast. Second, there was a focus on quantitative measures and models of physical factors and energy flow. Third, field experiments elucidated biological interactions.

Phytoplankton and benthic algae dominate primary production in kelp beds. Most of the production from macroalgae is not consumed by grazers, but is fragmented by waves, releasing POM and DOM. Bacteria play a key role in remineralization and in metabolising the DOM and at least a portion of the POM. Bacteria also transform feces so that they can profitably be consumed by filter feeders. Phytoplankton, bacteria, POM, and feces are all consumed by filter feeders, which constitute 73%–75% of animal biomass and production.

Physical factors have a dominant influence, with wind intensity and direction controlling upwelling and hence the supply of nutrients. Day-to-day, seasonal, and interannual variations in productivity are all attributable to the effects of winds coupled with solar input. Productivity peaks in summer when solar input is maximal and southeasterly winds prevail, favoring upwelling. Upwelling also dictates imports and exports. During upwelling the system is at its most productive, but water is shifted offshore, exporting POM and denying filter feeders access to phytoplankton. During downwelling, productivity diminishes because nutrients decline, but POM and phytoplankton are returned to the kelp bed. Paradoxically, food is most available during winter when productivity is lowest.

Superimposed on the bottom-up effects of upwelling, there are strong top-down effects due to predation. The rock lobster *Jasus lalandii* has substantial effects on community composition. Where rock lobsters are abundant, mussels are depleted and their size composition altered, and urchins and other herbivores are decimated. This causes ripple effects that permeate through the food web. Foliar algae proliferate, encrusting corallines diminish, and juvenile abalone that normally shelter beneath urchins virtually disappear. Spatial differences and temporal changes in rock lobster populations therefore have profound consequences for the ecosystem.

Human influences add to the complexity. Fishing is the most significant human activity, and has substantially depleted populations of rock lobsters, abalone, and linefish. The ultimate consequences of this are still difficult to assess because of cascade effects through the ecosystem, and the existence of indirect and nontrophic effects. Solutions to human effects include the development of operational management procedures, promotion of an ecosystem perspective, involvement of users in co-management, and the implementation of marine protected areas.

One of the first papers published on the Kelp Bed Programme was entitled "Sun, waves, seaweeds and lobsters: the dynamics of a west cast kelp-bed" (Field et al. 1977). The title was remarkably prescient. With perhaps the addition of the word "wind," it captures all the major elements of the system.

References

Anderson, R.J., P. Carrick, G.J. Levitt, and A. Share. 1997. Holdfasts of adult kelp *Ecklonia maxima* provide refuges from grazing for recruitment of juvenile kelps. Marine Ecology Progress Series **159**:265–273.

Andrews, W.R.H., and L. Hutchings. 1980. Upwelling in the southern Benguela Current. Progress in Oceanography **9**:1–81.

Attwood, C.G., and B.A. Bennett. 1995. Modelling the effects of marine reserves on the recreational shore-fishery of the South-Western Cape, South Africa. South African Journal of Marine Science **16**:227–240.

Awad, A.A., C.L. Griffiths, and J.K. Turpie. 2002. Distribution of South African marine benthic invertebrates applied to the selection of priority conservation areas. Diversity and Distributions **8**:129–145.

Barkai, A., and G.M. Branch. 1988a. Energy requirements for a dense population of rock lobsters *Jasus lalandii*: novel importance of unorthodox food sources. Marine Ecology Progress Series **5**:83–96.

Barkai, A., and G.M. Branch. 1988b. The influence of predation and substratal complexity on recruitment and settlement plates: a test of the theory of alternative states. Journal of Experimental Marine Biology and Ecology **124**:215–237.

Barkai, A., and G.M. Branch. 1989. Growth and mortality of the mussels *Choromytilus meridionalis* (Krauss) and *Aulacomya ater* (Molina) as indicators of biotic conditions. Journal of Molluscan Studies **55**:329–342.

Barkai, R., and C.L. Griffiths. 1986. Diet of the South African abalone *Haliotis midae*. South African Journal of Marine Science **4**:37–44.

Barkai, A., and C.D. McQuaid. 1988. Predator-prey role reversal in a marine benthic ecosystem. Science **242**:62–64.

Borchers, P., and J.G. Field. 1981. The effects of kelp shading on phytoplankton production. Botanica Marina **24**:89–91.

Branch, G.M. 1984. Changes in populations of intertidal and shallow water communities in South Africa during the 1982–83 temperature anomaly. South African Journal of Science **80**:61–65.

Branch, G.M., and C.L. Griffiths. 1988. The Benguela Ecosystem. Part V. The coastal zone. Oceanography and Marine Biology Annual Review **26**:395–486.

Branch, G.M., and C.N. Steffani. 2004. Can we predict the effects of alien species? A case-history of the invasion of South Africa by *Mytilus galloprovincialis* (Lamarck). Journal of Experimental Marine Biology and Ecology. **300**:189–215.

Branch, G.M., A. Barkai, P.A.R. Hockey, and L. Hutchings. 1987. Biological interactions: causes or effects of variability in the Benguela ecosystem? South African Journal of Marine Science **5**:425–445.

Branch, G.M., R.C. Thompson, T.P. Crowe, J.C. Castilla, O. Langmead, and S.J. Hawkins. 2008. Rocky intertidal shores: prognosis for the future. In press *in* N.V.C. Polunin, editor, Aquatic ecosystems: trends and prospects. Cambridge University Press, Cambridge, Mass.

Brown, A.C., and N. Jarman. 1978. Coastal marine habitats. Pages 1241–1277 *in* M.J.A. Werger, editor, Biogeography and ecology of Southern Africa. W. Junk, The Hague.

Bustamante, R.H., and G.M. Branch. 1996. The dependence of intertidal consumers on kelp-derived organic matter on the West Coast of South Africa. Journal of Experimental Marine Biology and Ecology **196**:1–28.

Bustamante, R.H., G.M. Branch, S. Eekhout, B. Robertson, P. Zoutendyk, M. Schleyer, A. Dye, D. Keats, M. Jurd, and C.D. McQuaid. 1995a. Gradients of intertidal productivity around the coast of South Africa and their relationship with consumer biomass. Oecologia **102**:189–201.

Bustamante, R.H., G.M. Branch, and S. Eekhout. 1995b. Maintenance of an exceptional grazer biomass in South: subsidy by subtidal kelps. Ecology **76**:2314–2329.

Buxton, C.D., and M.J. Smale. 1989. Abundance and distribution patterns of three temperate marine reef fish (Teleostei: Sparidae) in exploited and unexploited areas off the Southern Cape Coast. Journal of Applied Ecology **26**:441–451.

Carter, R. 1982. Phytoplankton biomass and production in a southern Benguela kelp bed ecosystem. Marine Ecology Progress Series **8**:9–14.

Castilla, J.C., G.M. Branch, and A. Barkai. 1994. Exploitation of two critical predators: the gastropod *Concholepas concholepas* and the rock lobster *Jasus lalandii*. Pages 101–130 *in* R.W. Siegfried, editor, Rocky shores: exploitation in Chile and South Africa. Springer-Verlag, Ecological Studies, Berlin.

Chapman, P., and L.V. Shannon. 1985. The Benguela Ecosystem. Part II. Chemistry and related processes. Oceanography and Marine Biology Annual Review **23**:183–251.

Cochrane, K.L., C.J. Augustyn, A.C. Cockcroft, J.H.M. David, M.H. Griffiths, J.C. Groeneveld, M.R. Lipinski, M.J. Smale, and R.J.Q. Tarr. 2004. An ecosystem approach to fisheries in the southern Benguela context. African Journal of Marine Science **26**:9–35.

Cockcroft, A.C. 2001. *Jasus lalandii* "walkouts" or mass strandings in South Africa during the 1990s: an overview. Marine and Freshwater Research **52**:1085–1094.

Connell, J.H., and W.P. Sousa. 1983. On the evidence needed to judge ecological stability or persistence. American Naturalist **121**:789–824.

Crawford, R.J.M., L.V. Shannon, and D.E. Pollock. 1987. The Benguela ecosystem, Part IV. The major fish and invertebrate resources. Oceanography and Marine Biology Annual Review **25**:353–505.

Day, E., and G.M. Branch. 2000a. Evidence for a positive relationship between juvenile abalone and the sea urchin *Parechinus angulosus* in the south-western Cape, South Africa. South African Journal of Marine Science **22**:137–144.

Day, E., and G.M. Branch. 2000b. Relationships between recruits of the abalone *Haliotis midae*, encrusting corallines and the sea urchin *Parechinus angulosus*. South African Journal of Marine Science **22**:145–156.

Day, E., and G.M. Branch. 2002a. Effects of sea urchins (*Parechinus angulosus*) on juveniles and recruits of abalone (*Haliotis midae*). Ecological Monographs **72**:133–149.

Day, E.G., and G.M. Branch. 2002b. Influences of the sea urchin *Parechinus angulosus* on the feeding behaviour and activity rhythms of juveniles of the South African abalone *Haliotis midae*. Journal of Experimental Marine Biology and Ecology **276**:1–17.

Dieckmann, G.S. 1980. Aspects of the ecology of *Laminaria pallida* (Grev.) J. Ag. off the Cape Peninsula (South Africa). I. Seasonal growth. Botanica Marina **23**:579–585.

Emanuel, B.P., R.H. Bustamante, G.M. Branch, S. Eekhout, and F.J. Odendaal. 1992. A zoogeographic and functional approach to the selection of marine reserves on the West Coast of South Africa. South African Journal of Marine Science **12**:341–354.

Field, J.G., and C.L. Griffiths. 1991. Littoral and sublittoral ecosystems of southern Africa. Pages 323–346 *in* A.C. Mathieson and P.H. Niehaus, editors, Ecosystems of the world **24**. Elsevier, Amsterdam.

Field, G.J., N.G. Jarman, G.S. Dieckmann, C.L. Griffiths, B. Velimirov, and P. Zoutendyk. 1977. Sun, waves, seaweed and lobsters: the dynamics of a west coast kelp-bed. South African Journal of Science **73**:7–10.

Field, J.G., C.L. Griffiths, R.J. Griffiths, N. Jarman, P. Zoutendyk, B. Velimirov, and A. Bowes. 1980. Variation in structure and biomass of kelp bed communities along the south-west Cape Coast. Transactions of the Royal Society of South Africa **44**:145–203.

Glassom, D., K. Prochazka, and G.M. Branch. 1997. Short-term effects of an oil spill on the West Coast of the Cape Peninsula, South Africa. Journal of Coastal Conservation **3**:155–168.

Griffiths, C.L. 2000a. Overview on current problems and future risks. Pages 235–241 *in* G. Preston, G. Brown, E. van Wyk, editors, Best management practices for preventing and controlling invasive alien species. Working For Water Programme, Cape Town.

Griffiths, M. 2000b. Long-term trends in catch and effort of commercial linefish of South Africa's Cape Province. Snap-shots of the 20th century. South African Journal of Marine Science **22**:81–110.

Griffiths, C.L., and J.L. Seiderer. 1980. Rock-lobsters and mussels—limitations and preferences in a predator-prey interaction. Journal of Experimental Marine Biology and Ecology **44**:95–109.

Harris, J.M., G.M. Branch, B.M. Clark, A. Cockcroft, C. Coetzee, A. Dye, M. Hauck, A. Johnson, L. Kati-Kati, Z. Maseko, K. Salo, W. Sauer, N. Siqwana-Ndulo, and M. Sowman. 2002. Recommendations for the management of subsistence fisheries in South Africa. South African Journal of Marine Science **24**:503–523.

Harris, J.M., G.M. Branch, C. Sibiya, and C. Bill, C. 2003. The Sokhulu subsistence mussel-harvesting project: co-management in action. Pages 61–98 *in* M. Hauck and M. Sowman, editors, Waves of change: coastal and fisheries co-management in South Africa. University of Cape Town Press, Cape Town.

Hauck, M., and N.A. Sweijd. 1999. A case study of abalone poaching in South Africa and its impact on fisheries management. ICES Journal of Marine Science **56**:1024–1032.

Hockey, P.A.R., and G.M. Branch. 1997. Criteria, objectives and methodology for evaluating marine protected areas in South Africa. South African Journal of Marine Science **18**:369–383.

Jerardino, A., and R. Navarro. 2002. Cape rock lobster (*Jasus lalandii*) remains from South African west coast shell middens: preservational factors and possible bias. Journal of Archaeological Science **29**:993–999.

Johnston, S.J., and D.S. Butterworth. 2005. Evolution of operational management procedures for the South Africa West Coast rock lobster (*Jasus lalandii*) fishery. New Zealand Journal of Marine and Freshwater Research **39**:687–702.

Koop, K., R.A. Carter, and R.C. Newell. 1982. Mannitol-fermenting bacteria as evidence for export from kelp beds. Limnology and Oceanography **27**:950–954.

Koop, K., and C.L. Griffiths. 1982. The relative significance of bacteria, meio- and macro-fauna on an exposed sandy beach. Marine Biology **66**:295–300.

Kuris A.M., and C.S. Culver. 1999. An introduced sabellid polychaete pest infesting cultured abalones and its potential spread to other California gastropods. Journal of Invertebrate Biology **118**:391–403.

Levitt, G.J., R.J. Anderson, C.J.T. Boothroyd, and F.A. Kemp. 2002. The effects of kelp harvesting on its regrowth and the understorey benthic community structure at Danger Point, South Africa, and a new method of harvesting kelp fronds. South African Journal of Marine Science **24**:71–85.

Matthews, S.G., and G.C. Pitcher. 1996. Worst recorded marine mortality on the South African coast. Pages 89–92 *in* T. Yasumoto, Y. Oshima and Y. Fukuyo editors, Harmful and toxic algal blooms. Intergovernmental Oceanographic Commission of UNESCO, Paris.

Mayfield, S., and G.M. Branch. 2000. Inter-relationships among rock lobsters, sea urchins and juvenile abalone: implications for community management. Canadian Journal of Fisheries and Aquatic Science **57**:2175–2187.

Mayfield, S., G.M. Branch, and A.C. Cockcroft. 2000. Relationships among diet, growth rate and food availability for the South African rock lobster, *Jasus lalandii*. Crustaceana **73**:812–834.

Mayfield, S., E. de Beer, and G.M. Branch. 2001. Prey preference and consumption of sea urchins and juvenile abalone by captive rock lobsters (*Jasus lalandii*). Marine and Freshwater Research **52**:773–780.

Newell, R.C., and J.G. Field. 1983. The contribution of bacteria and detritus to carbon and nitrogen flow in a benthic community. Marine Biology Letters **4**:23–36.

Newell, R.C., M.I. Lucas, B. Velimirov, and L.J. Seiderer. 1980. Quantitative significance of dissolved organic losses following fragmentation of kelp (*Ecklonia maxima* and *Laminaria pallida*). Marine Ecology Progress Series **2**:45–59.

Newell, R.C., J.G. Field, and C.L Griffiths. 1982. Energy balance and significance of micro-organisms in a kelp bed community. Marine Ecology Progress Series **8**:103–113.

Newman, G.G., and D.E. Pollock. 1974. Growth of the rock lobster *Jasus lalandii* and its relationship to benthos. Marine Biology **24**:339–346.

Pitcher, G.C. 1999. Harmful algal blooms of the Benguela current. Sea Fisheries Research Institute, Cape Town.

Pollock, D.E. 1979. Predator-prey relationships between the rock lobster *Jasus lalandii* and the mussel *Aulacomya ater* at Robben Island on the Cape West Coast of Africa. Marine Biology **52**:347–356.

Pollock, D.E. 1986. Review of the fishery for and biology of the Cape rock lobster *Jasus lalandii* with notes on larval recruitment. Canadian Journal of Fisheries and Aquatic Sciences **43**:2107–2117.

Pollock, D.E., and L.V. Shannon. 1987. Response of rock-lobster populations in the Benguela ecosystem to environmental change—a hypothesis. South African Journal of Marine Science **5**:887–899.

Pulfrich, A., C.A. Parkins, and G.M. Branch. 2003a. The effects of shore-based diamond-diving on intertidal and subtidal biological communities and rock lobsters in southern Namibia. Aquatic Conservation: Marine and Freshwater Ecosystems **13**:233–255.

Pulfrich, A., C.A. Parkins, G.M. Branch, R.H. Bustamante, and C.R. Velasquez. 2003b. The effects of sediment deposits from Namibian diamond mines on intertidal and subtidal reefs and rock lobster populations. Aquatic Conservation: Marine and Freshwater Ecosystems **13**:257–278.

Roberts, C.M., S. Andelman, G.M. Branch, R. Bustamante, J.C. Castilla, J. Dugan, B. Halpern, K.D. Lafferty, H. Leslie, J. Lubchenco, D. McArdle, H. Possingham, M. Rucklehaus, and R. Warner. 2003a. Ecological criteria for evaluating candidate sites for marine reserves. Ecological Applications **13**:Supplement S199-S214.

Roberts, C.M., G.M. Branch, R. Bustamante, J.C. Castilla, J. Dugan, B. Halpern, K.D. Lafferty, H. Leslie, J. Lubchenco, D. McArdle, M. Rucklehaus, and R. Warner. 2003b. Application of ecological criteria in selecting marine reserves and developing reserve networks. Ecological Applications **13**:Supplement S215–S228.

Russouw, J., and M. Russouw. 1999. Re-evaluation of recommended design wave method. Pages 486–498 *in* G.P. Mocke, editor, Fifth international conference on coastal and port engineering in developing countries. Council for Scientific and Industrial Research, Pretoria.

Seiderer, L.J., and R.C. Newell. 1985. Relative significance of phytoplankton, bacteria and plant detritus as carbon and nitrogen resources for the kelp bed filter-feeder *Choromytilus meridionalis*. Marine Ecology Progress Series **22**:127–139.

Seiderer, L.J., and F.T. Robb. 1986. Adaptive features of a bacteriolytic enzyme from the style of the mussel *Choromytilus meridionalis* in response to environmental fluctuations. Actes de Colloques **3**:427–433.

Seiderer, L.J., B.D. Hahn, and L. Lawrence. 1982. Rock-lobsters, mussels and man: a mathematical model. Ecological Modelling **17**:225–241.

Shafir, A., and J.G. Field. 1980. Importance of a small carnivorous isopod in energy transfer. Marine Ecology Progress Series **3**:203–215.

Shannon, L.V. 1985. The Benguela Ecosystem. Part I. Evolution of the Benguela, physical features and processes. Oceanography and Marine Biology Annual Review **23**:105–182.

Shannon, L.V., and S.C. Pillar. 1987. The Benguela ecosystem, Part III. Plankton. Oceanography and Marine Biology Annual Review **24**:65–179.

Sink, K.J., G.M. Branch, and J.M. Harris. 2005. Biogeographic patterns in rocky intertidal communities in KwaZulu-Natal, South Africa. African Journal of Marine Science **27**:81–96.

Stuart, V. 1982. Absorbed ration, respiratory costs and resultant scope for growth in the mussel *Aulacomya ater* (Molina) fed on a diet of kelp detritus of different ages. Marine Biology Letters **3**:289–306.

Stuart, V., M.I. Lucas, and R.C. Newell. 1981. Heterotrophic utilisation of particulate matter from the kelp *Laminaria pallida*. Marine Ecology Progress Series **4**:337–348.

Stuart, V., J.G. Field, and R.C. Newell. 1982a. Evidence for absorption of kelp detritus by the ribbed mussel *Aulacomya ater* using a new ^{51}Cr-labelled microsphere technique. Marine Ecology Progress Series **9**:263–271.

Stuart, V., R.C. Newell, and M.I. Lucas. 1982b. Conversion of kelp debris and faecal material from the mussel *Aulacomya ater* by marine micro-organisms. Marine Ecology Progress Series **7**:47–57.

Tarr, R.J.Q., P.V.G. Williams, and A.J. McKenzie. 1996. Abalone, sea urchins and rock lobster: a possible ecological shift that may affect traditional fisheries. South African Journal of Marine Science **17**:319–323.

Tugwell, S., and G.M. Branch. 1989. Differential polyphenolic distribution among tissues in the kelps *Ecklonia maxima, Laminaria pallida* and *Macrocystis angustifolia* in relation to plant defense theory. Journal of Experimental Marine Biology and Ecology **129**:219–230.

Tugwell, S., and G.M. Branch. 1992. Effects of herbivore gut surfactants on kelp polyphenol defenses. Ecology **72**:205–215.

Velimirov, B., and C.L. Griffiths. 1979. Wave-induced kelp movement and its importance for community structure. Botanica Marina **22**:169–172.

Velimirov, B., J.G. Field, C.L. Griffiths, and P. Zoutendyk. 1977. The ecology of kelp bed communities in the Benguela upwelling system. Analysis of biomass and spatial distribution. Helgolander wissenschaftliche Meeresunters-suchungen **30**:495–518.

Walter, R.C., R.T. Buffer, N.J. Bruggemann, M.M.M. Guillaume, S.M. Berhe, B. Negassi, Y. Libsekal, H. Cheng, R.L. Edwards, R. von Cosel, D. Néraudeau, and M. Gagnon. 2000. Early human occupation of the Red Sea coast of Eritrea during the last interglacial. Nature **405**:65–69.

Wulff, F.V., and J.G. Field. 1983. Importance of different trophic pathways in a nearshore benthic community under upwelling and downwelling conditions. Marine Ecology Progress Series **12**:217–228.

4

Subtidal Kelp-Associated Communities off the Temperate Chilean Coast

José M. Fariña, Alvaro T. Palma, and F. Patricio Ojeda

The Chilean coast is about 4,200 km long and can be divided into four main regions (fig 4.1; Santelices 1991, Castilla et al. 1993, Fernández et al. 2000):

1. Northern arid coast: extending from 18° to 27°S, characterized by inland arid climatic conditions and by a highly productive upwelling-dominated coastal ocean. From a geomorphological perspective this region shows a fairly straight coastline with few sheltered bays.
2. Semiarid coast: from 27° to 32°S, characterized by Mediterranean climate conditions with rainfalls occurring mostly during winter and with drought periods during summer. Oceanographically, this region combines high seasonal thermal variations with localized upwelling centers. The coastline is also fairly straight with only minor river inputs.
3. Central coast: from 32° to 42°S, also characterized by Mediterranean conditions. In this region, several rivers reach the coast and the occurrence of sheltered bays, as well as localized upwelling centers are common.
4. Fjords coast: from 43° to 56°S, characterized by the presence of an oceanic climate with heavy rains throughout the year.

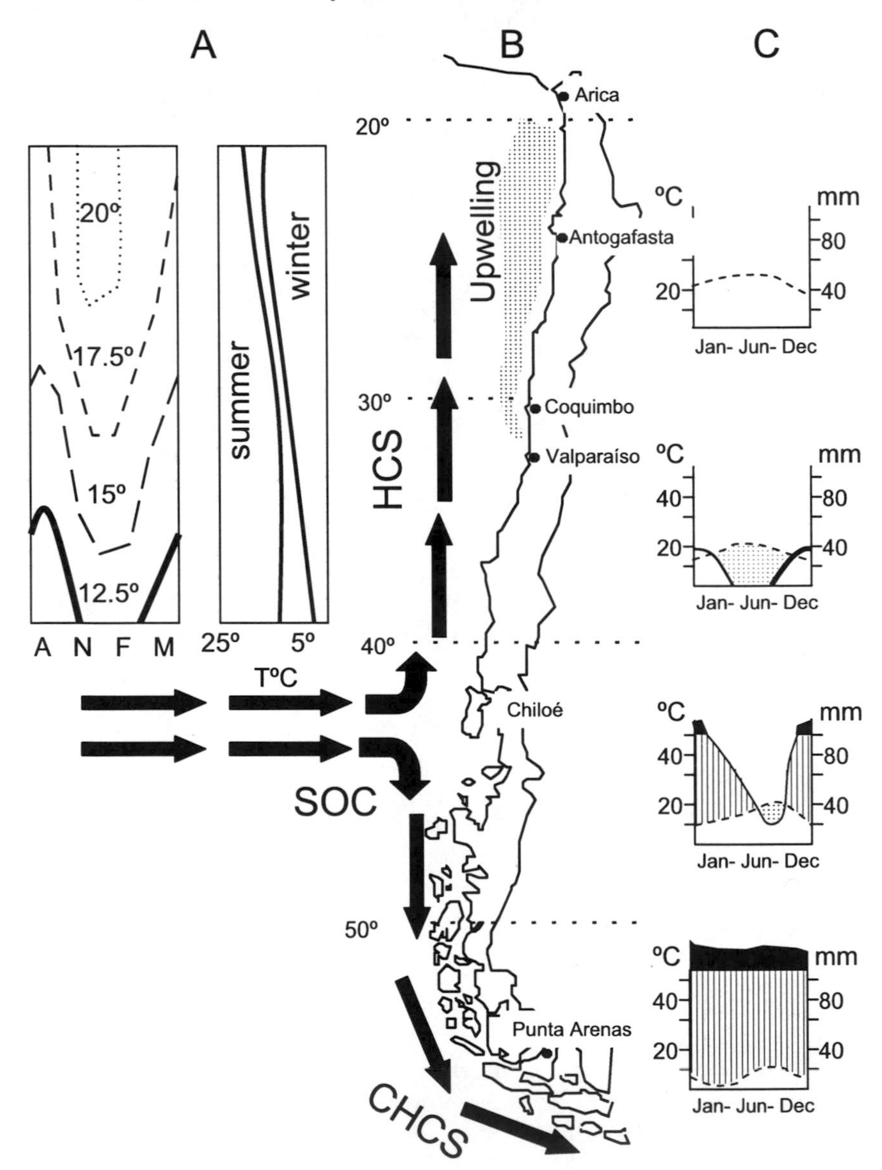

FIGURE 4.1 Main geographical regions, marine currents, upwelling areas, and main cities along the Chilean coast. HCS, Humboldt Current System; SOC, Southern Ocean Current; CHCS, Cape Horn Current System.

The shoreline is highly indented dominated by islands and fjords. The average seawater temperature is the lowest along the Chilean coast.

The first three regions described above are under the general influence of the Humboldt Current System (HCS), while the southern

Fjord coast is under the influence of the Cape Horn Current System (CHCS) (Escribano et al. 2002, Fernández et al. 2000).

Coastal Oceanography

The Southern Ocean Current (SOC) reaches Pacific South America near 45°S latitude, branching into two unequal masses of water flowing in opposite directions (fig. 4.1, Santelices 1991, Ahumada et al. 2000, Fernández et al. 2000, Escribano et al. 2002). Deflected to the south, the Cape Horn Current (CHC) follows the southern coast of Chile, rounds the tip of the continent projecting itself through the Drake Passage. The northwardly deflected mass, the Humboldt Current (HC), flows along the west coast of South America up to 4°27'S (northern Perú) where it turns toward the northwest across the Pacific Ocean. The HC flows 300 to 400 km off the coast and is only one of several currents and counter currents found north of 40°S. Three main water masses—the Subantarctic Water Mass (SAAW), the Subtropical Surface Water Mass (STW), and the Ecuatorial Subsurface Water Mass (ESSW)—dominate the oceanographic regime of the HCS (Escribano et al. 2002).

The SAAW originates in the eastern branch of the anticyclonic gyre of the southeastern Pacific and flows northward. South of 35°S, 1000 km off the South American coast, it represents the dominant element throughout the year, determining temperatures of 10°C to 18°C and salinities of 32 to 34°/$_{00}$ in the uppermost 100 m of the water column. The STW flows 30 m below the surface (above the SAAW) with a southward direction with temperature and salinity over 18°C and 34.9°/$_{00}$, respectively. The southern limit of this current is located around 25–27°S during summer and 23–25°S during winter (Bernal et al. 1982). The ESSW originates in the Subsurface Counter Current called the Cromwell Current, which flows eastward. This poleward flowing current accumulates as subsurface water in the northern coast of Perú.

Knowledge related to the ESSW is probably the most important element in understanding the upwelling events affecting the Chilean coast and other coastal process. The low oxygen content of ESSW ($O_2 < 1$ ml^{-1}) results from remineralization of organic matter falling from the euphotic zone and reaching neutral buoyancy at the density of this water mass (Daneri et al. 2000). This process produces high nutrient concentrations (some of the highest values reported in the ocean), high levels of dissolved organic matter, high levels of trace metals (bioaccumulation is 10^3 over the water value), and favors the occurrence of anoxic processes (such as dissimilative denitrification; Daneri et al. 2000, Iriarte et al. 2000). Sea-surface temperature data indicate that upwelling intensifies during spring and early summer and decreases during winter and fall seasons (Rodríguez et al. 1991).

The previous description highlights the main features that depict the Chilean coast, at least the central and northern section, being among the most productive marine systems in the world (Daneri et al. 2000). In summary, this system is dominated by recurrent wind-driven upwelling conditions that promote high levels of new primary production (Barber and Smith 1981), which is mostly restricted to a narrow band along the coast (Iriarte et al. 2000, Wieters et al. 2003). Due to the interaction between this productive band and geographic features, most of the sheltered bays along the Central Coast (from 32° to 42°) correspond to the habitats of conspicuous and diverse communities of subtidal organisms known as embayment ecosystems (Bernal and Ahumada 1985).

El Niño and La Niña Effects

Oceanographic patterns affecting the northern coast of Chilean are substantially modified during El Niño-Southern Oscillation (ENSO) events (Iriarte et al. 2000). The HCS is invaded by a massive southerly penetration of warm oceanic subtropical and equatorial subsurface water (Camus 1990). Increased surface temperatures are accompanied by heavy rains along the coastal range and, among other oceanographic changes, by reduced upwelling and increased downwelling close to shore, producing catastrophic effects on some fish and coastal biota (Camus 1990). The frequency and severity, as well as the southward extension of this phenomenon, varies in an unpredictable way. During strong ENSO events abnormally high temperatures have been recorded as far as 35°S (Jordan 1991).

El Niño is brought about by intensification and relaxation of easterlies or southeast trade winds. The Southern Oscillation has been associated with the variation of atmospheric pressure system in the Pacific Ocean: the high pressure at the Pacific anticyclone and the low pressure of the Indonesia cyclone system. A weakening of the Pacific anticyclone produces a warm event known as the negative face of the Southern Oscillation, or ENSO. The magnitude of the south-east trade relaxation, and its timing in relation to "normal" conditions determines the strength of the resulting El Niño event (Jordan 1991).

The effects of El Niño events on the Chilean coast are varied, depending on their intensity, and are often perceived as "positive" or "negative" according to their effect on human populations or economic activities. Likewise, these effects can also be classified as biological or physical (Ahumada et al. 2000). Several physical alterations have been recognized (Arntz and Fahrbach 1996): (a) warming of the surface ocean waters and coastal waters (changes in temperature up to +6°C along the coast); the positive temperature anomaly decreases poleward; (b) warming of the air masses up to 2°C or 3°C over the average annual temperature; (c) deepening of the thermocline; (d) weakening of upwelling events; (e) intrusion towards the coast of the STW, producing a series of ecological changes; and (f) a rise of sea level up to 30 cm.

Among the biological effects: (a) migration of adults and death of juvenile and larval fish species; (b) mortality of some frondose algal species due to nutrient depletion, with the consequential habitat loss for several associated species; (c) bathymetric migration of certain coastal fish in order to reach appropriate temperature conditions; (d) decrease of sea lion population due to emigration or death of their food resources; and (e) increased growth or development of several opportunistic species, some of them of economic importance, including *Octopus vulgaris* and *Argopecten purpuratus*. Generally, these biological effects can be described as breeding or recruitment failures and abundance fluctuations of varying extent (Arntz and Fahrbach 1996). One of the most dramatic effects occurs when mass mortality finally leads to local extinction in some particular species, altering their geographical population structure or even their latitudinal distribution ranges. Some species do not always recover to their pre-El Niño conditions, and the occurrence of a new event can have unpredictable consequences if coupled with prior ones. To date, the real ecological or evolutionary effect of El Niño on coastal species or communities, and especially on subtidal marine communities, remains largely unknown. However, as reviewed below, some studies show important changes on the trophic behavior of some species and predict alterations on their geographical distribution (Peters and Breeman 1993, Vargas et al. 1999).

The Food Web

Different biogeographical studies agree with respect to the major distributional patterns of Chilean biota along the coast (Santelices 1991, Camus 2001). Despite the fact that names and exact boundaries vary, depending on the researcher and data utilized, three major biogeographical units, or broad provinces of temperate nature, have been recognized: (a) The Peruvian province, a northern warm-temperate region ranging from Perú to latitude 30°S; (b) the Magellanic province, a southern cold-temperate region extending from latitude 42°S down to 56°S; and (c) an intermediate region between these two regions (30° to 42°S).

Large-sized kelps are the most representative species of the Chilean Littoral seascape. Between 18°S and 42°S, *Lessonia trabeculata* in the subtidal and *Lessonia nigrescens* in the intertidal, are the most conspicuous species (Santelices 1991, Camus and Ojeda 1992). South of 42°S, following the increase in representation of protected habitats, the dominants species are *Macrocystis pyryfera* in the subtidal, and *Durvillea antarctica* in the intertidal.

Most of the trophic analyses carried out along the northern and central Chilean coast (18° to 42°S) have focused on communities inhabiting rocky intertidal habitats-located in or above the *Lessonia nigrescens* belt (Castilla 1981, Moreno and Jaramillo 1983, Fariña et al. 2003). For the southernmost region (south of 42°S), subtidal communities associated with *Macrocystis pyrifera* forests were analyzed in a series of studies developed during the late 1970s and early 1980s (Dayton et al. 1977, Moreno and Jara 1984, Dayton 1985b). In the Intermediate Region, between 30° and 42°S, subtidal protected embayment areas are usually dominated by dense kelp beds of *Lessonia trabeculata*, which support diverse and conspicuous communities of consumers (Villouta and Santelices 1984).

Many studies have analyzed the basic ecological aspects of these communities, but a synthesis focused on their tropho-dynamic aspects is still lacking. Our general aim, therefore, is to review the available information relevant to this topic for the Chilean temperate shallow marine ecosystems, incorporating our own data in order to generate a basic food web consisting of the most important species inhabiting these systems. Basic to this scheme is discussion on: (1) the contrast in food webs between kelp-bed areas versus areas without kelps; (2) the possible transformation caused by ENSOs; and (3) the large effects of human exploitation.

Tropho-dynamic Studies on Subtidal Communities

Jaksic (1997) synthesized the history of the ecological studies on Chilean subtidal communities. The first systematic descriptions were developed during the mid 1900s during the expedition of the Lund University to Chile (1948–1950). Prof. Nibaldo Bahamonde, one of the founders of Chilean marine biology, published the results of this survey in a series of articles (Bahamonde 1950a,b; 1951a,b; 1952; 1953a,b,c,d; 1954; 1956; 1958; Bahamonde and Carcamo 1959; Henríquez and Bahamonde 1964), popularizing a methodology that was the base for later analyses. These works summarized important information on the natural history of several fish species (Moreno 1981). Later, with the development of industrial fisheries, several studies analyzed the trophic behavior of commercial fish species (Moreno 1981) and related behavior with parameters such as trophic web position (Hulot and Hermosilla 1960, Movillo and Bahamonde 1971), morphology and morphometrics (Moreno 1971, Ojeda 1986), trophic status (Moreno 1972, 1980; Ojeda and Camus 1977), trophic niche and bathymetric distribution (Moreno and Osorio 1977), prey availability and predation risk (Duarte and Moreno 1981), morpho-functional adaptations and restrictions to different diets (Cancino et al. 1985, Benavides et al. 1986, Fuentes and Cancino 1990, Vial and Ojeda 1990, 1992; Aldunate and de la Hoz 1993), and ontogenic changes on their dietary behavior (Benavides 1990, Benavides et al. 1994a,b).

From this period, three studies developed by Moreno and collaborators (Moreno et al. 1979, Moreno 1981, Moreno and Jara 1984) synthesized important ecological information. Moreno and colleagues (1979) reported an increase in the number of subtidal fish species along the Chilean coast from South (Antarctic) to North (Tropical) areas, largely due to an increase of herbivorous species. Specifically: (a) there are no strict herbivorous fishes in Antarctica; (b) at Valdivia (39°S) there are no strict herbivores but some omnivore species; (c) at San Antonio (33°S) one herbivorous species (*Aplodactylus punctatus*) appears; and (d) from La Serena (29°S) to Arica (18°S) there are many herbivore species (mainly Kyphosidae); and (e) the Peruvian and Ecuatorian subtidal communities are dominated by herbivorous species (mainly Percoidea). These authors associated this pattern with the increase in macroalgae species richness and abundance observed through the same geographical range. They also mention that algae directly serve as food for herbivores and indirectly increase the abundance of carnivorous fishes by providing refugia and by increasing habitat complexity for mostly invertebrate prey. The later observation is very important for the understanding of the dynamic and structure of subtidal marine communities living on the embayment ecosystems of the central Chilean coast (from 30° to 42°S).

Moreno (1981) highlighted the importance of frondose algal species as shelter for fishes. This author demonstrated that bathymetric segregation of subtidal fishes is related to the type of substratum and habitat heterogeneity and not to prey distribution. This pattern was associated with fish escape behavior due to predation exerted by mobile and big-sized vertebrates, such as sea lions. Furthermore, Moreno and Jara (1984) demonstrated a close relationship between fish species composition and habitat structure of subtidal areas. Several species occur exclusively in kelp areas while others were restricted to boulders, barren-grounds, and sand bottom areas. This pattern was also correlated with the feeding behavior of fishes.

During the past decade several studies continued the tradition of describing the natural history of subtidal fish species (Fariña and Ojeda 1993, Ojeda and Fariña 1996, Palma and Ojeda 2002) along with studies on the feeding ecology of these organisms (Cáceres et al. 1993, Vargas et al. 1999, Angel and Ojeda 2001). Most of these studies highlight the importance of habitat complexity and structure (associated with the presence of kelp) as a major determinant of prey availability, which determines consumer fish preferences.

In spite of the considerable amount of information contained in the above-mentioned studies, the energetic dynamic of Chilean rocky subtidal communities is practically unknown. However, several steady-state trophic flow models for benthic communities living in sea grass, sand-gravel, sand, and muddy subtidal benthic habitats have been published (Ortiz 2001, Ortiz and Wolff 2002), foretelling a promising line of work for future studies on the topic.

As to this point, it is clear that most of the tropho-dynamic studies developed on subtidal communities of central Chile focused on fishes as main consumers. A few exceptions show that benthic herbivores, such as the sea urchin *Tetrapygus niger*, omnivorous, such as the gastropod *Tegula tridentate*, and carnivores, such as the starfish *Meyenaster gelatinosus*, exert an important influence on kelps and their associated fauna distribution (Vasquez 1993). However, those effects are restricted to shallow subtidal areas (between 0 and 5 m deep). This situation could be related with the preferential use of drift algae and detritus rather than fresh algae by benthic invertebrates on deeper (5–15 m) subtidal areas (Rodriguez 2000). There are no studies concerning the possible effect of large-bodied vertebrates, including sea lions or sea birds, on subtidal communities and those organisms are anecdotally mentioned in the later studies.

Food Web Components

For 20 years we have been studying the components of subtidal communities of the Chilean coast, from Antarctica to Iquique at the northern region. In this chapter we will synthesize, from a trophic web perspective, the information regarding subtidal ecosystems of the central coast (between 32° and 40°S). We first review the general information available for each trophic level and then present the results of our own studies. These results are based on surveys and dietary analyses of benthic invertebrates and algae and demersal or bentho-pelagic fishes living in close association with subtidal kelp beds occurring in four embayments within the above-mentioned geographical range.

Detailed descriptions of the methods utilized can be found elsewhere (Camus and Ojeda 1992, Cáceres et al. 1993, Ojeda and Fariña 1996, Angel and Ojeda 2001, Palma and Ojeda 2002). In short, fishes were sampled during the four seasons in each embayment using experimental gillnets (with graded mesh from 10 to 70 mm) randomly set at depths between 5 and 20 m. Visual counts of fish were also carried out in order to complement the use of gillnets. Benthic algae and invertebrates were also seasonally sampled from 12 plots of 0.25 m^2 randomly located at approximately 10 m of depth along the bottom of each embayment. Captured specimens were fixed in formalin and transported to the laboratory for further analysis. The abundance of captured fish species was determined by a catch-per-unit-effort (CPUE) index corresponding to the total number of captured individuals of each species divided by the total number of sampling hours during the study period. Fish biomass was determined multiplying the average number of individual for each species observed on the visual surveys by average biomass of each species from gillnets. Benthic organism's biomass was directly recorded from the sampled plots. In the laboratory, specimens were identified to species level, measured in size and wet-weighed. When possible, all the digestive tracts were removed and their prey items contents identified to the lowest possible taxonomic level, damp-dried on a paper towel, and weighed. Dietary composition was expressed as the percentage of total food weight pooled over all individuals for each fish species.

Each consumer species was classified as belonging to herbivores, phytoplanktivore-detritivore, or carnivore in accordance with the biomass of their percentage dietary composition of animals, detritus, plankton, or algae. Species were classified in those categories if the consumption of the respective prey (algae, plankton-detritus or animals) was 80% of the total biomass that they consumed. These results allowed us to estimate species composition, abundance, biomass and their status within the trophic web.

Primary producers Marine communities are supported by many sources of organic matter, including benthic algae and phytoplankton. As exemplified by many ecological studies (Moreno et al. 1979, Moreno 1981, Moreno and Jara 1984, Vargas et al. 1999, Angel and Ojeda 2001), it is mainly large algae that support Chilean rocky subtidal communities and phytoplankton plays only a secondary role as primary producer. Virtually unknown is the role of bacteria and other microorganisms as component of these communities.

Rocky subtidal habitats of the central and northern Chilean coast are dominated by the brown algae *Lesssonia trabeculata*, which forms kelp forests (*sensu* Steneck et al. 2002) of variable size on stable rocky substratum from shallow to 20-m depth (Villouta and Santelices 1984). On semi-protected sites, the outer edge of the kelp can reach the limit between subtidal and intertidal zones but, in the case of more exposed areas, the kelp formation is located at around 2-m depth (Santelices 1989). Normally, the deeper limits of these kelps are associated with substratum discontinuities and their coastal extension varies between 500 to 100 m depending on the inclination of the substratum. Plant density varies between and within sites ranging from 4 to 0.5 plants/m^2 and it is inversely associated with substratum discontinuities, inclination and depth. Holdfast diameter and plant biomass usually shows a bathymetric pattern of variation. In shallow areas (≈2-m depth) holdfast diameters are around 18 cm and plants biomass is around 6 kg/m^2. At intermediate depths (≈5 m) holdfast size and plant biomass reach their maxima at 24 cm and 10 kg/m^2, respectively while at around 15-m depth both variables decrease to values of around 12 cm and 1 kg/m^2 respectively. It has been reported that at shallow depths plants density shows an inverse relationship with herbivore density, which mostly include sea urchins and gastropods (Vasquez 1993).

As shown in similar systems, an important part of kelp production can be exported out of subtidal landscapes into intertidal and deeper waters where it energetically subsidizes filter and detritus feeders (Duggins et al. 1989). In Chile, it was originally proposed that due to morphological constrains, the most abundant subtidal benthic herbivore, the black urchin *Tetrapygus niger*, was unable to capture and use drift algae as a food source (Vasquez 1986, Contreras and Castilla 1987). Later, Vasquez (1993) proposed that aggregating could solve this constraint, and drift algae have since been shown to be a common food for this urchin (Rodriguez 1999, Rodriguez and Fariña 2001).

Our surveys have shown that understory species are mostly dominated by crustose calcifying but also by some other non-calcifying fleshy algae of the genera *Glossophora, Plocamium, Zonaria, Bossiella, Rhodymenia, Gelidium* and *Halopteris*. Under *L. trabeculata* kelps, crustose calcifying algae, small sized algae, sessile invertebrates, and fishes are common, showing bathymetric differences in their species compositions and dominances. Between 5- to 7-m depths, crustose algae and tubicolous polychaeta dominate the substratum, while crustose noncalcifying algae (*Hildenbrandtia* and *Peysonella*), bryozoa and barnacles occur at lower abundances. At greater depths of between 17 and 19 m other fleshy algae, such as *Bossiella orbignata, Gelidium* and *Halopteris*, increase their cover while small sponges and bryozoa are uncommon.

A total of 24 species of alga have been recorded in our surveys on subtidal areas of central Chile (tab. 4.1). In terms of biomass, brown algae are composed of four species and are the dominant group with values near 8 kg/m^2. Red algae are composed of 15 species and show intermediate values of around 65 g/m^2, while green algae include five species and are the least abundant group, at around 5 g/m^2. In terms of species biomass, the substratum is dominated by *Lessonia trabeculata* (8800 g/m^2) followed by several species of red algae such as *Gelidium chilense* (40 g/m^2), *Hildebrandtia lecannellieri* (20 g/m^2), *Plocamium* sp. (20 g/m^2) and *Glossophora* sp. (15 g/m^2). In concordance with previous studies, green algal species, described as herbivore's preferred food, were represented in our surveys mostly by *Ulva* sp., at low biomasses of around 1.5 g/m^2.

Consumers In our surveys, 60 species of benthic invertebrates were recorded and mollusks and crustaceans were the most diverse groups at 30 and 20 species, respectively. In terms of the biomass of the 24 most representative species (tab. 4.1), two invertebrates

Table 4.1. Main trophic and functional groups of Chilean subtidal communities associated with *Lessonia trabeculata* kelps.

Gross Trophic Level	Functional Group	Dominant Taxa	Biomass (g/m²)
Primary producers	Phytoplankton	Non-ident.	28.0
	Brown alga (4 spp.)	*Lessonia trabeculata*	8805.1
	Green alga (5 spp.)	*Ulva* sp.	1.5
	Red alga (15 spp.)	*Gelidium chilense*	44.8
		Hildenbrandtia lecannellieri	22.7
		Plocamium sp.	21.2
		Glosophora sp.	13.9
		Crustosa calcarea	11.4
		Rhodomenia sp.	5.6
		Acrochaetium sp.	3.5
Planktivores and detritivores	Zooplankton	Non-ident.	18.0
	Polychaetes	Non-ident.	9.9
		Phragmatopoma sp.	4.6
	Bivalves	*Protothaca thaca*	13.6
		Chama pellucida	9.3
	Barnacles	*Balanus laevis*	7.1
	Tunicates	*Pyura chilensis*	25.0
	Fishes	*Chromis crusma*	10.6
		Isacia conceptionis	27.2
Herbivores	Chitons	*Tonicia* sp.	5.2
		Chiton cumingsii	8.3
	Gastropods	*Turritela* sp.	7.6
		Tegula atra	12.5
		Tegula cuadricostata	6.5
		Tegula tridentata	84.5
		Prisogaster niger	14.0
		Mitrella sp.	1.8
		Fisurella latimarginata	2.1
		Crepipatela dilatata	2.8
		Nucella crassilabrum	5.0
		Nassarius sp.	16.5
	Amphipods	Gamaridae	0.9
	Crabs	*Pisoides edwardsi*	3.2
		Pagurus comptus	3.3
	Urchins	*Tetrapygus niger*	215.9
	Fishes	*Aplodactylus punctatus*	47.5
Carnivores	Gastropods	*Concholepas concholepas*	41.9
	Crabs	*Cancer setosus*	162.4
		Homalaspis plana	6.7
	Fishes	*Schroederichthys chilensis*	2.0

(Continued)

Table 4.1. (Continued)

Gross Trophic Level	Functional Group	Dominant Taxa	Biomass (g/m²)
		Sebastes capensis	10.1
		Cheilodactylus variegatus	15.2
		Pinguipes chilensis	65.6
		Bovichthys chilensis	65.6
	Sea lions	*Otaria flavescens*	10.0

Biomass corresponds to the wet weight of the most important species belonging to each functional group.

dominated; the urchin *Tetrapygus niger* (215 g/m^2) and the snail *Tegula tridentata* (84 g/m^2), followed by a carnivore, the gastropod *Concholepas concholepas* (41 g/m^2) and by the filter-feeder tunicate *Pyura chilensis* (25 g/m^2).

Subtidal fish assemblages of the central Chilean coast are usually represented by several carnivore species (*Cheilodactylus variegatus, Mugiloides chilensis, Sebastes capensis, Semicossyphus maculatus* and *Graus nigra*) and few herbivores (*Aplodactylus punctatus*). However, in terms of biomass, herbivores are the dominant group. In our surveys a total of 24 fish species were identified (fig. 4.2). The most representative species of this group (tab. 4.1) were two planktivores-detritivores, one herbivore, and five carnivores (tab. 4.1). In terms of biomass, the carnivores (pooled together 103 g/m^2) dominate the assemblage followed by the unique herbivore, *Aplodactylus punctatus* (47.5 g/m^2). All those fish species have been recorded to be the food source of some marine mammals such as *Otaria byronia*, which occurs occasionally on protected embayments for less than a month (Moreno et al. 1979). It is worth mentioning that just during the last five years the previously rare Chilean sea otter *Lutra felina* has been observed on subtidal environments of the central Chilean coast (A. Palma personal observation). Crabs (54%), fishes (40%), and gastropods (9.5%) are the principle components of the sea otter diet (Ostfeld et al. 1989).

Food Web Diagram

From a general diagram containing the feeding relationships and biomass it is clear that more than 99% of the ecosystem's total biomass occurs on the benthos (fig. 4.3). This is due to the high incidence of brown algae and because the presence of the large demersal consumers, such as sea lions and fishes are less common. Most of the benthic consumers are invertebrates, with a high incidence of herbivorous species. Within this group, mollusks and echinoderms are dominant. In the water column, however, fishes represent the dominant

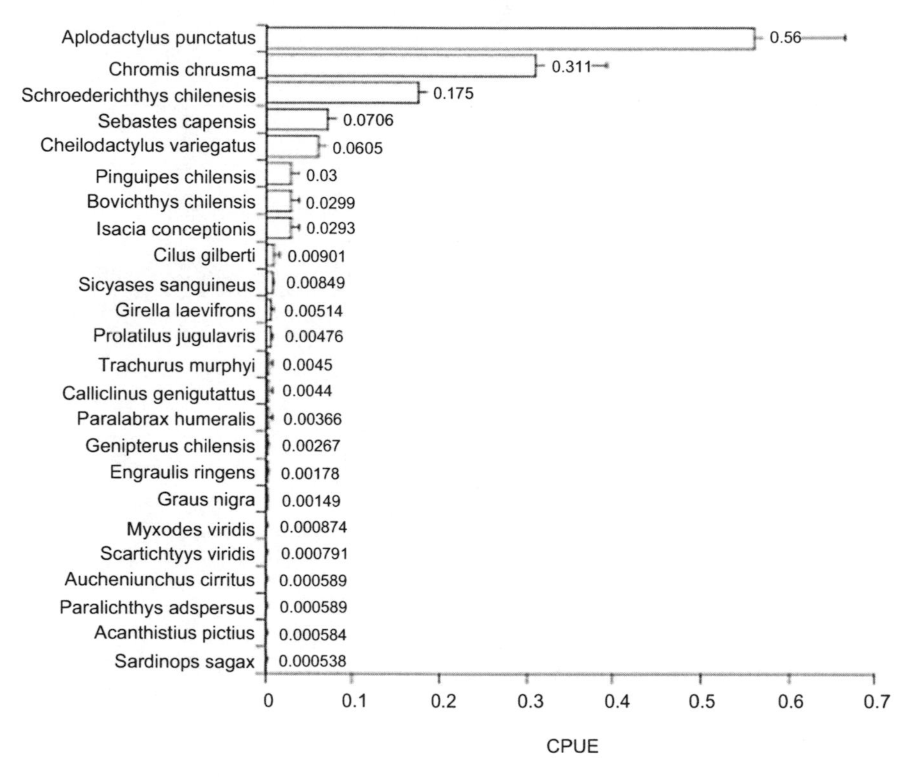

FIGURE 4.2 Relative abundances, measured as Capture Per Unit Effort (CPUE in individuals/hour) of the pelagic fish species captured in our surveys with a gill net.

group, mainly composed of carnivores followed by one herbivore species and by several planktivores (fig. 4.2).

In the case of invertebrates, omnivory is widespread, while in the case of demersal species; the group of planktivores fishes (composed by *Chromis crusma* and *Isacia conceptionis*) depends exclusively on planktonic organisms and has no trophic connections with the benthic zone. All the rest of the consumers depend directly or indirectly on benthic prey and ultimately on algae as a primary producer. Because the top predators are highly specialized carnivorous fishes, the structure of the trophic network presents four trophic levels (fig. 4.3).

The trophic relationships of the 67 consumer species recognized in our surveys (59 benthic invertebrates and six demersal or benthopelagic fishes) were categorized in 20 functional groups (tab. 4.1) belonging to the above four main trophic levels. Demersal fish and benthic herbivores, such as chitons and gastropods, consume primary producers, which also represents the main energy source for consumers of detritus, which are mainly crabs. Phytoplankton, on the other hand, is an important source for a reduced group of filter

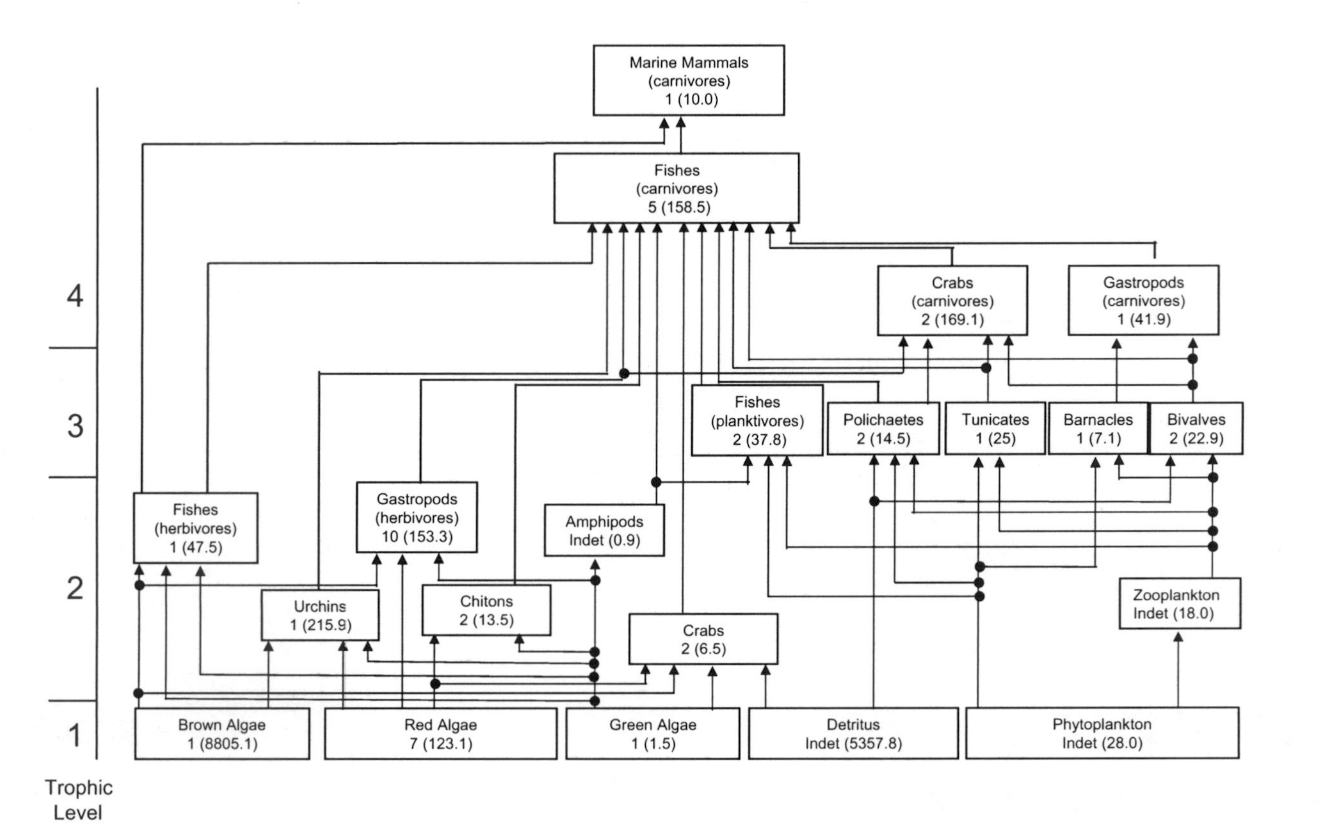

FIGURE 4.3 General trophic web diagram of the subtidal communities associated with kelp forest on the central Chilean coast. For each box component the number of species and total biomass is shown.

feeders, such as bivalves and barnacles, as well as for zooplankton. However, a more detailed description on the trophic role of phytoplankton in these communities is still unknown. Likewise, resuspended detritus represents a source of energy for the same filter feeders in this system.

A suit of diverse organisms that feed mostly on primary production composes the second and third trophic levels of our ecosystem. However, a portion of them are mainly supported by benthic algae (left side of diagram in fig. 4.3) and directly feed on a few top-level carnivorous fish consumers. The other first-level consumers, mostly maintained by detritus and phytoplankton, are consumed by fish but also by carnivorous crabs. The later organisms represent an intermediate level in this web, since they prey upon several invertebrates at the third trophic level, but are only preyed upon by fish. The top trophic level, represented here by mammals (sea lions), is supported by several fish species.

Temporal and Spatial Variation

There are no studies describing the spatial and temporal variation of subtidal communities as a whole but some studies make detailed descriptions of the temporal variation in abundance of some of the above species from specific central Chilean sites. Three studies specifically describe how temporal (Vargas et al. 1999) and spatial (Nuñez and Vasquez 1987, Angel and Ojeda 2001) aspects of habitat complexity affect the trophic ecology of subtidal fishes.

Nuñez and Vasquez (1987) studied the trophic amplitude and microhabitat use by four sympatric fish species inhabiting the subtidal kelp forest of the northern Chilean coast (at 28°S). Their results show a vertical segregation of the species, associated to some degree with their feeding behavior. The strict herbivore, *Aplodactylus punctatus,* occurred in close association with the kelp forest; the planktivorous and carnivorous dominant species, *Chromis crusma* and *Cheilodactylus variegatus*, were respectively recorded in the water column, while the other carnivorous species, *Pinguipes chilensis*, occurred mostly close to the bottom. The authors concluded that these species had low microhabitat and dietary overlap and a strong relationship between their food and their spatial distribution.

Angel and Ojeda (2001) compared the feeding ecology of 34 species of subtidal fishes living in two localities of the northern Chilean coast (around 27°S) with differences in kelp forest cover. In the presence of kelp forest, both carnivores and omnivores consumed mainly benthic prey inhabiting the understory canopy, while in its absence, carnivores fed on prey living in the water column. These authors concluded that a more complex spatial structure would seem to support a more complex trophic organization with a greater degree of connectivity and interaction between species in the ecosystem.

Vargas and colleagues (1999) analyzed the temporal variation of the trophic ecology of seven subtidal fish species living in the northern areas (around 20°S). Their results showed that, for a period of nine years, most of the species maintained their functional trophic status (defined as herbivore, carnivore, and omnivore) but changed dramatically their prey species-specific preferences. These changes were explained by the changes in the frondose algal populations associated with large-scale perturbations, such as ENSO.

It has been reported that kelp deforestation occurs during El Niño events. As explained already, El Niño halts coastal upwelling of nutrient-rich water and causes a warming of the surface waters and this produces patchy deforestation, which is usually followed by a rapid recovery (Dayton et al. 1989). The physiological stress is likely to be more common toward the low latitudinal limit of kelp ranges. Under such conditions, it has been shown that the limit of three common brown algae shifted towards higher latitudes following the El Niño event of 1982–1983 (Peters and Breeman 1993). Because subtidal communities of the temperate Chilean coast depend mostly on kelps for both food and habitat (Steneck et al. 2002), it is possible to predict that the whole subtidal community and their trophic relations will change during ENSO years. Until now, however, there is no study analyzing the changes.

From both the analysis of the trophic web structure and the information related to the temporal variation of Chilean subtidal communities, it is evident that the dominant kelp species plays a major role as primary producer, but also as habitat for most of the consumer species. Consequently, *Lessonia trabeculata* is a foundational species (sensu Dayton 1985a).

Human Influences

Artisanal Fisheries

The most important human influences on rocky subtidal zones off the Chilean coast are the collection of invertebrates, fish, and large frondose algae by artisanal fishers and occasional coastal gatherers. Thus, humans as components of this ecosystem can impose significant alterations on population and community structure (Castilla 1999).

Direct and indirect effects The fauna of edible benthic invertebrates is extremely rich and diverse, comprising about 60 species, of which a high percentage is unique and geographically restricted to the

Chilean or Southeastern Pacific coastline (Castilla 1994). Among these, gastropod mollusks, crabs, sea urchins, sea squirts, and a giant barnacle are the most heavily exploited along the Chilean coast (Castilla and Defeo 2001).

Collection of invertebrates and frondose algae for human use can be traced back to hunter-gatherers (9,000 years ago) in northern and central Chile (Jerardino et al. 1992). The introduction of diving technology and the access to new foreign markets over the past few decades has, however, produced an exponential increase in exploitation rates, which in 2004 was 432,587 annual tons, and has led to symptoms of overexploitation of the main target species (Fernández et al. 2000). Although it is well known that many of the species collected have the potential to play critical roles in these communities as predators or basal and structural species, the direct ecological effects of their removal are still unknown. From the wide diversity of species collected by humans, those whose collection has potentially the most wide-spread consequences for the rest of the community are the carnivore muricid gastropod *Concholepas concholepas*, the herbivore keyhole limpets of the genus *Fissurella*, the sea urchin *Loxechinus albus*, and large carnivorous fishes such as the kyphosid *Graus nigra* and the labrid *Semicossyphus maculatus*, (Fernández et al. 2000, Ojeda et al. 2000). Further, kelp drift has been shown to be an important food for many invertebrate species, particularly for sea urchins (Castilla and Moreno 1982, Rodriguez 2003). Nonetheless, human exploitation of these species should influence the structure and functioning of these rocky subtidal communities: directly, by altering their population sizes; and indirectly, by affecting food web relationships. Most of these species are either important prey, such as crabs, mollusks, and barnacles, or predators, such as *Concholepas concholepas* and large fishes (Castilla 1981).

Space-providing species Kelps of the brown algae *Lessonia trabeculata* and *Macrocystis integrifolia*, beds of the sea squirt *Pyura chilensis*, and of mussels in the genus *Aulacomya* and *Chorumytilus* are the most important space-providing species occurring in subtidal reefs (Santelices 1991). These species are commonly and intensively harvested for the alginate industry and for human consumption. Kelp forests play a fundamental role as habitat providers in rocky subtidal environments (Dayton 1985a). Experimental studies with *Lessonia trabeculata* have revealed several ecological effects of harvesting. Either partial upper canopy or complete plant removal has negative consequences for both the kelp and invertebrate-associated fauna (Vasquez and Santelices 1990).

The most important population effect of kelp and invertebrate removal is the increased distance between plants, which increases access to kelp holdfasts and other surfaces by sea urchins and snail grazers. This reduces kelp recruitment and alters plant morphology, which makes them more susceptible to being removed by water movements (Vasquez and Santelices 1990). Angel and Ojeda (2001) found similar invertebrate species diversity and fish abundance when comparing two localities of the north of Chile that differed in kelp abundance but there were clear differences in species dominance, number of functional groups, and trophic guilds. There were also marked differences in the trophic webs between both localities. In the presence of kelp forest, both carnivores and omnivores consume mainly benthic prey living in the understory canopy while, in its absence; carnivores feed on prey present in the water column. The more complex spatial structure provided by kelp seems, therefore, to increase the complexity of trophic organization, adding a greater degree of connectedness as well as enhanced interactions between species.

Summary

The confluence of physical, geological, and oceanographic factors define the marine ecosystems along the Chilean coast. Within these ecosystems, subtidal kelps beds of shallow protected embayments are widely distributed along the central coast (30°–42°S). *Lessonia trabeculata*, the dominant algae on those kelp beds, can be categorized as a biogenic and basal foundation species because it is the main primary producer and also serves as habitat for several species of invertebrate and vertebrate consumers. The subtidal communities of the central Chilean coast depend on benthic algal productivity, are dominated by few consumer species, and show a high incidence of herbivores and omnivores. Many vertebrates and invertebrates that potentially play important roles as consumers or structural species are commercially exploited; however the ecological effects of their removal are still poorly understood. Kelps are also under commercial exploitation and its removal has similar negative consequences for both the kelp itself and its associated fauna. There is a chain of synergic effects associated with kelp removal. Removal increases the distance between plants, which increases the effect of grazers. The reduction in kelp increases the reliance of species on pelagic prey, which is considerably less productive than benthic production. This is expected to lead to decreased productivity of the ecosystem and associated fisheries.

References

Ahumada, R.B., L.A. Pinto, and P.A. Camus. 2000. The Chilean coast. Pages 669–717 *in* C.R.C. Sheppard, editor, Seas at the millennium: an environmental evaluation. Volume I, Regional Chapters: Europe, The Americas and West Africa. Pergamon, Amsterdam, The Netherlands.

Aldunate, R., and E. de la Hoz. 1993. Diversidad trófica de *Cheiron pisciculus* (Ostariophysi: Characidae):¿consequencia de una versatilidad del mecanismo alimentario?. Revista Chilena de Historia Natural **66**:177–184.

Angel, A., and F.P. Ojeda. 2001. Structure and trophic organization of subtidal fish assemblages on the northern Chilean coast: the effect of habitat complexity. Marine Ecology Progress Series **217**:81–91.

Arntz, W.E., and Fahrbach, E. 1996. El Niño: Experimento climático de la naturaleza. Fondo de Cultura Económica, Ciudad de México.

Bahamonde, N. 1950a. Alimentación del Peje-gallo (*Calloryinchus callorhynchus*). Investigaciones Zoológicas Chilenas **1**:4–6.

Bahamonde, N. 1950b. Alimentación del Rollizo (*Pinguipes chilensis*, Molina 1781). Investigaciones Zoológicas Chilenas **1**:12–14.

Bahamonde, N. 1951a. Alimentación del Tollo (*Squalus fernandinus*, Molina 1782). Investigaciones Zoológicas Chilenas **1**:9–10.

Bahamonde, N. 1951b. Alimentación de la Sierra (*Thyrsites atun*, (Euphrasen), (1791). Investigaciones Zoológicas Chilenas **1**:8–10.

Bahamonde, N. 1952. Alimentación de la Pintarroja (*Halaelurus chilensis* (Guichenot) 1948). Investigaciones Zoológicas Chilenas **1**:3–5.

Bahamonde, N. 1953a. Alimentación de la Huelca (*Macrohunus magellanicus* Lömmberg, 1907) Investigaciones Zoológicas Chilenas **1**:5–7.

Bahamonde, N. 1953b. Alimentación del Chancharro (*Helicolenus lengerichi* Norman, 1937). Investigaciones Zoológicas Chilenas **1**:8–9.

Bahamonde, N. 1953c. Alimentación del la Raya (*Raja flavirostris* Philippi, 1892). Investigaciones Zoológicas Chilenas **2**:7–8.

Bahamonde, N. 1953d. Alimentación de la Merluza de los canales (*Merluccius australis* (Hutton), 1872). Investigaciones Zoológicas Chilenas **2**:23–30.

Bahamonde, N. 1954. Alimentación de los Lenguados (*Paralichtys microps* e *Hipoglosina microps*). Investigaciones Zoológicas Chilenas **2**:72–74.

Bahamonde, N. 1956. Alimentación de la Palometa (*Cheilodactylus gayi* Kner). Investigaciones Zoológicas Chilenas **3**:29–30.

Bahamonde, N. 1958. Sobre el contenido estomacal de ejemplares de Merluza (*Merluccius gayi-gayi*) capturados en Coquimbo. Boletin Informativo del Departamente de Pezca y Caza **54**:9–12.

Bahamonde, N., and M. Carcamo. 1959. Obervaciones sobre la alimentación de la Merluza (Merluccius gayi) de Talcahuano. Investigaciones Zoológicas Chilenas **5**:211–216.

Barber, R.T., and R.L. Smith. 1981. Coastal upwelling ecosystems. Pages 31–68 *in* A.R. Longhurst, editor, Analysis of marine ecosystem. Academic Press, N.Y.

Benavides, A.G. 1990. Variación ontogenética de la capacidad para asimilar algas de *Aplodactylus punctatus* (Pisces: Aplodactylidae). Tésis de magister, Facultad de Ciencias, Universidad de Chile, Santiago.

Benavides, A.G., F. Bozinovic, J.M. Cancino, L. Yates. 1986. Asimilación de algas por dos peces del litoral chileno: *Syciases sanguineus* (Gobiescidae) y *Aplodactylus punctatus* (Aplodactylidae). Medio Ambiente (Chile) **8**:21–26.

Benavides, A.G., J.M. Cancino, and F. P. Ojeda. 1994a. Ontogenetic change in the diet of *Aplodactylus punctatus* (Pisces: Aplodactylidae): an ecophysiological explanation. Marine Biology **118**:1–5.

Benavides, A.G., J.M. Cancino, and F. P. Ojeda. 1994b. Ontogenic changes in gut dimensions and macroalgal digestibility in the marine herbivorous fish, *Aplodactylus punctatus*. Functional Ecology **8**:46–51.

Bernal, P. and R.B. Ahumada. 1985. Ambiente Oceánico. Pages 55–106 *in* F. Soler, editor, Medio Ambiente en Chile. Centro de investigación y planificación del Medio Ambiente (CIPMA). Ediciones Universidad Católica, Santiago, Chile.

Cáceres, C.W., A.G. Benavides, and F.P. Ojeda. 1993. Ecología trófica del pez herbívoro *Aplodactylus punctatus* (Pisces: Aplodactlylidae) en la costa centro-norte de Chile. Revista Chilena de Historia Natural **66**:185–194.

Camus, P.A. 1990. Procesos regionales y fitogeografía en el Pacífico Sudeste: el efecto de "El Niño-Oscilación del Sur." Revista Chilena de Historia Natural **63**:11–17.

Camus, P.A. 2001. Biogeografía marina de Chile continental. Revista Chilena de Historia Natural **74**:587–617.

Camus, P.A., and F.P. Ojeda. 1992. Scale-dependent variability of density estimates and morphometric relationships in subtidal stands of the kelp *Lessonia trabeculata* in northern and central Chile. Marine Ecology Progress Series **90**:193–200.

Cancino, J.M., C.A. Moreno, and J. Garrido. 1985. Relaciones entre características morfológicas del tubo digestivo y dieta de peces litorales. Archivos de Biología y Medicina Esperimentales (Chile) **18**:R89.

Castilla, J.C. 1981. Perspectivas de investigación en estructura y dinámica de comunidades intermareales rocosas de Chile central. II. Depredadores de alto nivel trófico. Medio Ambiente (Chile) **5**:190–215.

Castilla, J.C. 1994. The Chilean small-scale benthic shellfisheries and the institutionalization of new management practices. Ecology International Bulletin **21**:47–63.

Castilla, J.C. 1999. Coastal marine communities: trends and perspectives from human-exclusion experiments. Trends in Ecology and Evolution **14**:280–283.

Castilla, J.C., and C.A. Moreno. 1982. Sea urchins and *Macrocystis pyrifera*: experimental test of their ecological relations in southern Chile. Pages 257–263 *in* J.M. Lawrence, editor, International Echinoderm Conference, Tampa Bay. A.A. Balkema, Rotterdam.

Castilla J.C., and O. Defeo. 2001. Latin-American benthic shellfisheries: emphasis on co-management and experimental practices. Reviews in Fish Biology and Fisheries **11**:1–30.

Castilla, J.C., S.A. Navarrete, and J. Lubchenco. 1993. Southern Pacific coastal environments: Main Features, large-scale perturbations, and global climate change. Pages 167–188 *in* Mooney, H., E. Fuentes, and B. Kronberg, editors, Earth system response to global change: contrasts between north and south America. Academic Press. N.Y.

Contreras, S., and J.C. Castilla. 1987. Feeding behaviour and morphological adaptation in two sympatric sea urchin species in central Chile. Marine Ecology Progress Series **38**:217–224.

Daneri, G., V. Dellarossa, R.A. Quiñones, B. Jacob, P. Montero, and O. Ulloa. 2000. Primary production and community respiration in the Humboldt Current System off Chile and associated oceanic areas. Marine Ecology Progress Series **197**:43–51.

Dayton, P.K. 1985a. Ecology of kelp communities. Annual Review Ecology and Systematics **16**:215–245.

Dayton, P.K. 1985b. The structure and regulation of some South American kelp communities. Ecological Monographs **55**:447–468.

Dayton, P.K., R.J. Rosenthal, L.C. Mahan, and T. Antezana. 1977. Population structure and foraging biology of the predaceous Chilean asteroid *Meyenaster gelatinosus* and the escape biology of its prey. Marine Biology **39**:361–370.

Dayton, P.K., Tegner, M.J., Edwards, P.B., and K.L Riser. 1989. Temporal and spatial scales of kelp demography. The role of oceanography and climate. Ecological Monographs **69**:219–250.

Duarte, W., and C.A. Moreno. 1981. The specialized diet of *Harpagifer bispinis*: its effect on the diversity of Antarctic intertidal amphipods. Hydrobiologia **81**:241–250.

Duggins, D.O., C.A. Simenstad, and J.A. Estes. 1989. Magnification of secondary production by kelp detritus in coastal marine ecosystems. Science **245**:170–173.

Escribano, R., V.H. Marín, P. Hidalgo and G. Olivares. 2002. Physical-biological interactions in the pelagic ecosystem of the nearshore zone of the northern Humboldt Current System. Pages 31–43 *in* J.C. Castilla, J.L. Largie, editors, The oceanography and ecology of the nearshore and bays in Chile. Proceedings of the international symposium on linkages and dynamics of coastal systems: open coasts and embayments, Santiago, Chile 2000. Ediciones Universidad Católica de Chile, Santiago, Chile.

Fariña, J.M., and F.P. Ojeda. 1993. Abundance, activity and trophic patterns of the redspotted catshark, *Schroederichthys chilensis,* in the Pacific temperate coast of South America. Copeia **1993**:545–549.

Fariña, J.M., J.C. Castilla, and F.P. Ojeda. 2003. On the relationship between species richness and productivity: the "idiosyncratic" effect of a "Sentinel" species on rocky intertidal communities in northern Chile affected by cooper mine tailings. Ecological Applications **13**:1533–1552.

Fernández, M., E. Jaramillo, P. Marquet, C. Moreno, S. Navarrete, F.P. Ojeda, C. Valdovinos, and J. Vásquez. 2000. Diversity, dynamics and biogeography of Chilean benthic nearshore ecosystems: an overview and guidelines for conservation. Revista Chilena de Historia Natural **73**:797–830.

Fuentes, L.S., and J.M. Cancino. 1990. Cambios morfométricos en el tubo digestivo de juveniles de *Girela laevifrons* (Kyphosidae) en función de la dieta y del nivel de repleción. Revista de Biología Marina (Chile) **25**:19–26.

Henríquez, G., and N. Bahamonde. 1964. Análisis cualitativo y cuantitativo del contenido gástrico del congrio negro (*Genypterus maculatus* (Tschudi) en pescas realizadas entre San Antonio y Constitución (1961–1962). Revista Universitaria (Universidad Católica de Chile) **49**:139–158.

Hulot, A., and I. Hermosilla. 1960. Posición de *Merluccius gayi-gayi* en la cadena alimenticia del Pacífico frente a la zona de Concepción (Chile). Actas del 1er Congreso Sudamericano de Zoología (La Plata, Argentina): 115–122.

Iriarte, J.L., G. Pizarro, V.A. Troncoso, and M. Sobarzo 2000. Primary production and biomass of size-fractionated phytoplankton off Antofagasta, Chile (23–24°S) during pre-El Niño and El Niño 1997. Journal of Marine Systems **26**:37–51.

Jaksic, F.M. 1997. Ecología de los Vertebrados de Chile. Ediciones Universidad Católica de Chile, Santiago de Chile.

Jerardino A., J.C. Castilla, J.M. Ramirez, and N. Hermosilla 1992. Early coastal subsistence patterns in central Chile: A systematic study on the marine-invertebrate fauna from the site of Curaumilla-1. Latin American Antiquity **3**:43–62.

Jordan, R.S. 1991. Impact of ENSO events on the southeastern Pacific region with special reference to the interaction of fishing and climate variability. Pages 401–430 *in* Glantz, M., Katz, R., and Nicholls, N., editors, ENSO teleconnections linking worldwide climate anomalies: scientific basis and societal impacts. Cambridge University Press, Cambridge, U.K.

Moreno, C.A. 1971. Somatometría y alimentación natural de *Harpagifer georgianus* antarcticus en Bahía Fildes, Isla Rey Jorge, Antártica. Boletín del Instituto Antártico Chileno **6**:9–12.

Moreno, C.A. 1972. Nicho alimentario de la "vieja negra" (*Graus nigra* Philippi) (Osteichthyes Labridae). Noticiero Mensual del Museo Nacional de Historia Natural (Chile) **186**:5–6.

Moreno, C.A. 1980. Observations on food and reproduction in *Trematomus barnacchii* (Pisces Nototheniidae) from Palmer Archipelago, Antarctica. Copeia **1980**:171–173.

Moreno, C.A. 1981. Desarrollo de los estudios sobre relaciones tróficas en peces del sublitoral rocoso antarctico y subantartico de Chile. Medio Ambiente **5**:161–174.

Moreno, C.A., and C. Osorio. 1977. Bathymetric food habits changes in the Antarctic fish *Notothenia gibberifrons* Lönberg (Pisces: Nototheniidae). Hydrobiologia **55**:139–144.

Moreno, C.A., W.E. Duarte, and J.H. Zamorano. 1979. Variación latitudinal del número de especies de peces sublitoral rocoso: una explicación ecológica. Archivos de Medicina y Biología Experimental **12**:169–178.

Moreno, C.A., and E. Jaramillo. 1983. The role of grazers in the zonation of intertidal macroalgae of the Chilean coast. Oikos **41**:73–76.

Moreno, C.A., and H.F. Jara. 1984. Ecological studies of fish fauna associated with *Macrosystis pyrifera* belts in the south of Fuegonian Islands, Chile. Marine Ecology Progress Series **15**:99–107.

Movillo, J., and N. Bahamonde. 1971. Contenido gástrico y relaciones tróficas de *Thyrsites atun* (Euphrasen) en San Antonio, Chile. Boletin del Museo de Historia Natural (Chile) **29**:289–338.

Nuñez, L.M., and J.A. Vasquez. 1987. Observaciones tróficas y de distribución especial de peces asociados a un bosque submareal de Lessonia trabeculata. Estudios Oceanológicos (Chile) **6**:79–85.

Ojeda, F.P. 1986. Morphological characterization of the alimentary tract of Antarctic fishes and its relation to feeding habits. Polar Biology **5**:125–128.

Ojeda, F.P., and J. Camus. 1977. Morfometría y nicho trófico de *Coelorhynchus patagoniae* Gilbert y Thompson. Boletín del Museo Nacional de Historia Natural (Chile) **35**:99–104.

Ojeda, F.P., and J.M. Fariña. 1996. Temporal variations of the abundance, activity and trophic patterns of the rockfish *Sebastes capensis*, off the central Chilean coast. Revista Chilena de Historia Natural **69**:205–211.

Ojeda, F.P., F. Labra, and A.A. Muñoz. 2000. Biogeographic patterns of Chilean littoral fishes. Revista Chilena de Historia Natural **73**:625–641.

Ortiz, M. 2001. Holistic modeling of a subtidal benthic ecosystem of northern Chile (Tongoy Bay), to improve the knowledge and understanding of its structure and function: assessing the effects of intensive fisheries upon different invertebrate and alage species. Dissertation Zur Erlangung de Grades Doktor de Naturwissenschaften, Universität Bremen.

Ortiz, M., and M. Wolff. 2002 Trophic models of four benthic communities in Tongoy Bay (Chile): comparative analysis and preliminary assessment of management strategies. Journal of Experimental Marine Biology and Ecology **268**:205–235.

Ostfeld, R.S., L. Ebensperger, L.L. Klosterman, and J.C. Castilla. 1989. Foraging, activity budget, and social behavior of the South American Marine Otter *Lutra felina* (Molina 1782). National Geographic Research **5**:422–438.

Palma, A.T., and F.P. Ojeda. 2002. Abundance distribution and feeding patterns of a temperate reef fish in subtidal environments of the Chilean coast: the importance of understory algal turf. Revista Chilena de Historia Natural **75**:189–200.

Peters, A.F., and A.M. Breeman 1993. Temperature tolerances and latitudial range of brown kelp from temperate Pacific South America, Marine Biology **115**:143–150.

Rodríguez, L.V. Marín, M. Farías, and E. Oyarce. 1991. Identification of an upwelling zone by remote sensing and in situ measurements. Mejillones del Sur Bay (Antofagasta, Chile). Scientia Marina **55**:467–473.

Rodriguez, S.R. 1999. Subsidios tróficos en ambientes marinos: la importancia de las macroalgas pardas a la deriva como fuente exógena de recursos para el erizo Tertapigus niger (Echinodermata: Echinoidea) en el intermareal rocoso de Chile Central. Tésis de Doctorado, Facultad de Ciencias Biológicas, Pontificia Universidad Católica de Chile, Santiago.

Rodriguez, S.R. 2000. Transferencia de recursos alimentarios entre diferentes ambientes del sistema costero. Revista Chilena de Historia Natural **73**:199–207.

Rodriguez, S.R. 2003. Consumption of drift kelp by intertidal populations of the sea urchin *Tertrapygus niger* on the central Chilean coast: possible consequences at different ecological levels. Marine Ecology Progress Series **251**:141–151.

Rodriguez, S.R., and J.M. Fariña. 2001. Effect of drift kelp on the spatial distribution pattern of the sea urchin *Tetrapygus niger* (Molina): geostatistical approach. Journal of the Marine Biological Association of the United Kingdom **81**:179–180.

Santelices, B. 1989. Comunidades de algas en el submareal rocoso de Chile continental, Capitulo V. Pages 79–91 *in* B. Santelices, Algas marinas de Chile. Distribución, ecología utilización y diversidad. Ediciones Universidad Católica de Chile, Santiago, Chile.

Santelices, B. 1991. Littoral and sublittoral communities of continental Chile. Pages 347–369 *in* A.C. Mathieson and P.H. Nienhuis, editors, Ecosytems of the world **24**. Intertidal and littoral ecosystems. Elsevier, Amsterdam, the Netherlands.

Steneck, R.S., Graham, M.H., Bourque, B.J., Corbett, D., Erlandson, J.M., Estes, J.A., and M.A. Tegner. 2002. Kelp forest ecosystems: biodiversity, stability, resilience and future. Environmental Conservation: **29**:436–459.

Vargas, M.E., R.A. Soto, and G.L. Guzmán. 1999. Cambios interanuales en la alimentación de peces submareales del norte de Chile entre los 20°11'S y 20°20'S. Revista de Biología Marina y Oceanografía **32**:197–210.

Vasquez, J.A. 1986. Morfología de estructuras alimentarias como factores en la organización de comunidades submareales. Biota (Chile) **1**:104.

Vasquez, J.A. 1993. Abundance, distributional patterns and diets of main herbivorous and carnivorous species associated to *Lessonia trabeculata* kelp beds in northern Chile. Serie Occasional **2**:213–229.

Vasquez, J.A., and B. Santelices. 1990. Ecological effects of harvesting *Lessonia* (Laminariales, Phaeophyta) in central Chile. Hydrobiologia **204/205**:41–47.

Vial, C.I., and F.P. Ojeda. 1990. Cephalic anatomy of the herbivorous fish *Girella laevifrons* (Osteichthyes: Kyphosidae): mechanical considerations of its trophic function. Revista Chilena de Historia Natural **63**:247–260.

Vial, C.I., and F.P. Ojeda. 1992. Comparative analysis of the head morphology of Pacific temperate kyphosid fishes: a morpho-functional approach to prey-capture mechanims. Revista Chilena de Historia Natural **65**:471–484.

Villouta, E., and B. Santelices 1984. Estructura de la comunidad submareal de *Lessonia* (Paheophyta, Laminariales) en Chile norte y central. Revista Chilena de Historia Natural **57**:111–122.

Wieters, E.A, D.M. Kaplan, S.A. Navarrete, A. Sotomayor, J. Largier, K.J. Nielsen, and F. Véliz. 2003. Alongshore and temporal variability in chlorophylla concentration in Chilean nearshore waters. Marine Ecology Progress Series **249**:93–105.

5

Diversity and Dynamics of Californian Subtidal Kelp Forests

Michael Graham, Ben Halpern, and Mark Carr

Californian kelp-forest ecosystems are highly diverse and productive and are one of the most distinctive features of the Californian coastline. They have also served as the focus of innumerable experimental and observational studies by ecologists interested in processes structuring nearshore marine systems. Despite over 50 years of intensive field and laboratory research, however, much remains to be understood about the processes that determine their diversity and dynamics. In particular, food-web structure of communities associated with the giant kelp (*Macrocystis pyrifera*) has received little attention (Rosenthal et al. 1974, Pearse and Hines 1976, Foster and Schiel 1985, Schiel and Foster 1986, Graham 2004), due in no small part to the apparently complex nature of these systems. Since Darwin (1839), it has been generally perceived that much of the structure and diversity of these ecosystems is due to the presence of giant kelp itself (Graham 2004). Yet, it is still unclear whether this role of giant kelp is due to its high levels of energy production, its provision of complex habitat, or if it is simply a by-product of the inherently diverse and productive coastal environments in which giant kelp is found. This chapter introduces the reader to the diversity and dynamics of Californian kelp forest food webs and explores the methodological and theoretical challenges of studying the processes structuring kelp forest communities.

Geological History

A diverse array of geological processes has shaped the Californian coastline (Legg 1991), establishing the regional physiography and geologic substrate composition. To begin with, the narrow continental shelf, deep ocean trenches, and steep mountain ranges that define much of California's coastal geomorphology are consequences of the subduction of various oceanic plates beneath the North American continental plate. Unlike most other regions of the world, however, this region was subsequently transformed into a strike-slip fault system, caused by the collision of an eastward migrating mid-ocean spreading center with the subduction zone, beginning approximately 40 million years ago (mya). The transformation was most dramatic south of Point Conception (fig. 5.1), where uplift, subsidence, rotation, compression and extension due to faulting and tectonics resulted in a mosaic of basins, islands and offshore banks embedded within a widened region of the continental shelf. Subduction ceases north of Point Conception, although the coastline remains relatively linear and the continental shelf relatively narrow. The composition of the substratum can vary over short spatial scales (<10 kilometers) in California, from basaltic, granitic, or sedimentary rocks to gravel and sandy beaches (Greene and Kennedy 1986, 1987, 1989). Extensive rocky platforms are more common north of Point Conception, whereas large sandy beaches are predominant to the south (Graham et al. 2003). Consequently, kelp forests north and south of Point Conception inhabit regions of fundamentally different geomorphology.

Northeast Pacific Oceanography

In California, coastal oceanographic parameters important to the establishment of kelp forests, such as temperature, nutrients, and wave action, are controlled primarily by variability in a cool southward-flowing eastern boundary current, the California Current, partially linked to swings in the Pacific Decadal Oscillation (Chelton and Davis 1982, Lynn and Simpson 1987, Lluch-Cota et al. 2001, Batchelder and Powell 2002, Bograd et al. 2003). During periods when the Aleutian low-pressure system is weak, spring equator-ward winds and the flow of the California Current are strong, and oceanographic productivity is fuelled by longshore transport of nutrients from upwelling regions at coastal promontories. As the Aleutian low strengthens or winds weaken during the fall, a poleward-flowing counter-current (the Davidson Current) often develops, bathing the coast in warmer, more nutrient-depleted waters. Consequently, seasonal fluctuations in sea-surface temperature, nutrients, and productivity are ubiquitous features of the Californian coast (Hickey 1998, fig. 5.2).

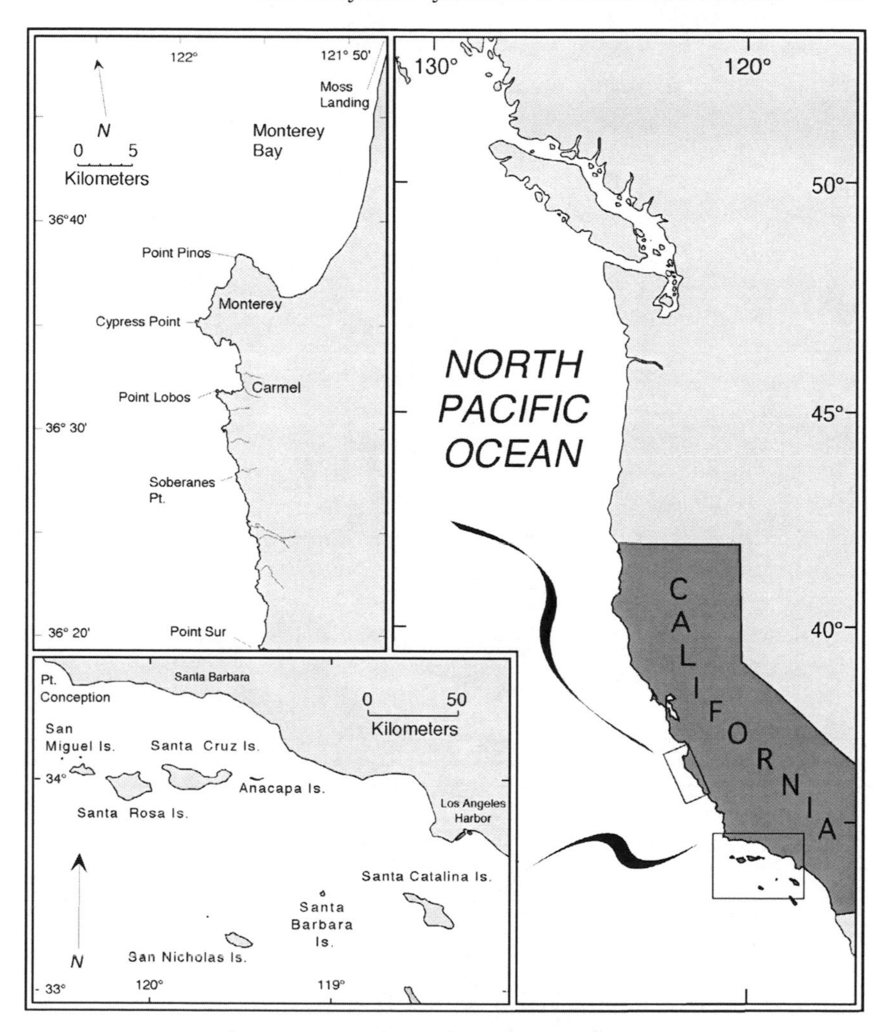

FIGURE 5.1 Location of central and southern Californian regions that include the kelp forests of primary focus in this paper. Point Conception marks the biogeographic transition point between the two regions, although the western Channel Islands (San Miguel and Santa Rosa) have fauna, flora, and oceanographic conditions similar to central California.

As with coastal geomorphology, however, differences in coastal oceanographic conditions can be striking north and south of Point Conception (Hickey 1998, Bograd et al. 2003). In southern California, a semi-permanent cyclonic gyre exists that incorporates California Current water with warmer waters intruding from the southeast (Hickey 1993). As such, seasonal fluctuations in oceanographic conditions are generally much greater south of Point Conception (fig. 5.2); for example, summer–fall sea-surface temperatures average >18°C in southern California but <15°C in central California. Greater distance

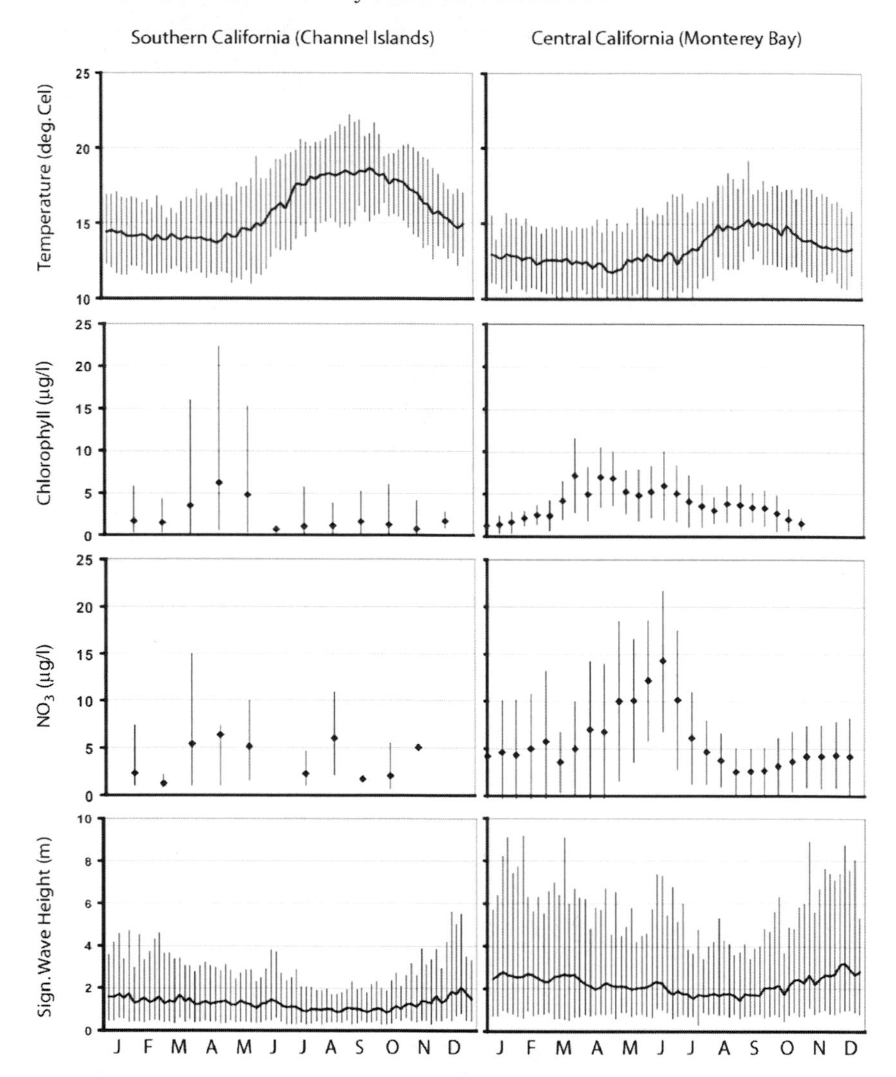

FIGURE 5.2 Environmental conditions for the Monterey Bay and Channel Islands regions. Dark lines and points are mean values; bars indicate the range of possible values for that date (week or month) across all years data were available, except for chlorophyll and nitrate in Monterey where bars are 1 SD. Temperature data come from 14 years of AHVVR satellite data (10 × 10 km grid size; July 1985–August 1999) and are 5-day averages of data from 10 grid cells each for the Monterey and Channel Islands regions. Significant wave height data are 5-day averages of hourly data from Buoy #46042 (Monterey; 1987–2002) and Buoy #46053 (Channel Islands, 1994–2002). Chlorophyll and nitrate data for Monterey Bay are from Mooring 1 (1990–2002); for the Channel Islands, the data are from CalCOFI tow stations 83.51, 83.42, 83.40.6, and 82.47 (1985–2002; means and ranges are calculated from all four tows).

from the Aleutian low, and increased protection by Point Conception and offshore islands, also result in a more benign wave environment in southern California (fig. 5.2). Finally, episodic El Niño-Southern Oscillation (ENSO) events (≈4–7 year periodicity) substantially alter oceanographic conditions in California by enhancing the poleward flow of warm waters, shutting down upwelling, raising nearshore sea levels, and increasing wave intensity (Ware and Thompson 2000). For kelp systems, each of these ENSO effects increases in severity from north to south (Edwards 2004), except wave intensity, which exhibits the opposite trend. These marked differences in the oceanographic conditions experienced by shallow coastal reefs north and south of Point Conception contribute to marked differences in the productivity and seasonal and interannual variability in kelp forest size and distribution (Edwards 2004).

Patterns of Biogeography and Biodiversity

The flora and fauna of Californian coastal waters are inherently rich, with current taxonomic resources describing approximately 650 macroalgal, 500 fish, 90 bird, 28 marine mammal, and thousands of invertebrate species (Miller and Lea 1972, Smith and Carlton 1975, Abbott and Hollenberg 1976, Austin 1985, Ricketts et al. 1968, Briggs et al. 1987, Orr and Helm 1989). Overall species composition and diversity of these taxonomic groups differ greatly between the coastal waters of central/northern and southern California. In general, the warmer Californian biogeographic province to the south of Point Conception is enriched in fish, invertebrate, and macroalgal taxa of sub-tropical origin, and has a higher rate of endemism, relative to the cooler Oregonian province north of Point Conception. The region around Point Conception is a transition zone between the two provinces and some taxonomic groups increase in diversity there due to provincial overlap and inclusion of "transition zone" taxa (Newman 1979); numerous other biogeographic boundaries have been proposed along the Californian coast (Dawson 2001). Many of these patterns in biodiversity and biogeography likely reflect historic and present patterns of geographic variability in geomorphology and oceanography, and it is within this physical gradient that Californian kelp forest communities have assembled and evolved.

The Food Web

Our objectives in constructing a general Californian kelp forest food web are three-fold. First, the food web presents an updated perspective on the general structure of trophic interactions within Californian kelp forests based on studies conducted since the last food web was constructed over 20 years ago (Foster and Schiel 1985). Second, in combination with a discussion of regional variation in common kelp forest taxa, the food web illustrates key similarities and differences in kelp forest associations between southern and central California. Finally, the patterns revealed from the food web can generate hypothesized roles of trophic interactions and habitat associations in structuring Californian kelp forest communities, and help guide process-oriented studies designed to test those hypotheses. In the first section (Trophic guilds), we (1) describe the members of each trophic level that create the food web and (2) introduce major differences in kelp forest species composition north and south of Point Conception. In the subsequent section (Food-web Structure) we evaluate key trophic and habitat associations and use the differences in such associations between southern and central kelp forests to suggest whether energy or habitat provision is structuring kelp forest systems, and to demonstrate possible experimental approaches that could be used to test these ideas.

Trophic Guilds

Producers The energy base of Californian kelp forests is founded primarily upon subtidal kelps (Order Laminariales) of variable productivity potential and habitat architecture (fig. 5.3). Of the six common subtidal kelp taxa in California, two species, *Macrocystis*

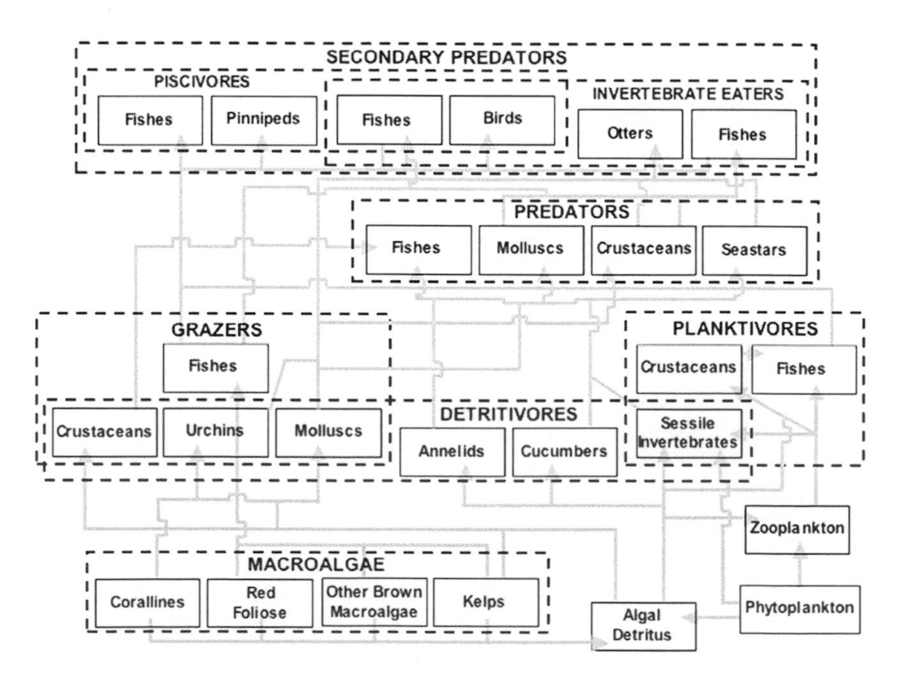

FIGURE 5.3 Californian kelp-forest food web for both central and southern California. Trophic groups are indicated by dashed boxes, dominant taxonomic groups within the trophic group are indicated with solid-line boxes, and trophic links are shown with directional arrows.

pyrifera and *Nereocystis luetkeana,* are dominant in terms of their contribution to the formation of surface canopies, productivity and habitat structure. A third species, *Pelagophycus porra,* is less common, typically forms a canopy below the surface, and is found only in southern California (Druehl 1970, Abbott and Hollenberg 1976). The giant kelp *M. pyrifera* dominates shallow nearshore rocky platforms (<25–30 m depth) in southern California and most regions of central California that range from moderately exposed to protected. The bull kelp *N. luetkeana* has a southern range limit to the north of Point Conception and is the primary canopy-former in areas of high wave exposure (Foster and Schiel 1985). Of the three other prominent subtidal kelps, *Pterygophora californica* and *Eisenia arborea* form dense canopies between 1 and 2 m above the reef surface, whereas blades of the more prostrate *Laminaria farlowii* lie across the surface of the reef; all three can be found both north and south of Point Conception. Together, these canopy, sub-canopy, and prostrate taxa form successive layers in Californian subtidal kelp forests, with some taxa found only in particular regions, habitats, or depth ranges (Spalding et al. 2003). Beneath these various kelp layers is an exceedingly diverse group of red, green, and brown foliose, turfing, and encrusting algae (Breda and Foster 1985, Harrold et al. 1988, Graham 2004). In addition to high coverage of benthic substratum, the foliose and turfing algae provide key habitat and energy resources to diverse epifaunal assemblages that can be important food sources for higher trophic levels, including fishes (Limbaugh 1955, Quast 1968a, Bray and Ebeling 1975, Coyer 1985, Hallacher and Roberts 1985, Ebeling and Laur 1986, Hobson and Chess 1986, Coyer 1987, Holbrook et. al. 1997).

No studies have directly compared non-kelp macroalgal assemblages between central and southern California, but regional studies have identified 43 species common to kelp forests in southern California (Graham 2004) and >50 species common to kelp forests along the Monterey Peninsula in central California (Harrold et al. 1988); only 25 species of macroalgae were shared between these two Californian studies. The primary difference in non-kelp macroalgal diversity between these regions appears to be due to a decrease in species richness of red algae, and an increase in brown algae (primarily of the orders Dictyotales and Fucales), from north to south (e.g., contrast Foster and Schiel 1992 and Graham 2004 with Breda and Foster 1985 and Harrold et al. 1988). Each of these algal groups also contributes to an algal detrital pool (fig. 5.3) that can be the primary conduit of fixed carbon into kelp systems (Gerard 1976, Pearse and Hines 1976, Harrold and Reed 1985, Duggins 1988, Duggins et al. 1989). Whether through attached or detrital pathways, it is generally accepted that, when present, *M. pyrifera* represents the single greatest source of fixed carbon to Californian kelp forests

(Graham 2004); it is also the primary structural component of the ecosystem.

Consumers The prominent consumers in Californian kelp forests constitute five trophic categories: grazers, detritivores, planktivores, and lower- and higher-level carnivores (fig. 5.3). The feeding habits of many consumer species defy strict categorization within a single feeding-category (Foster and Schiel 1985), as they acquire energy from a variety of trophic groups. For example, the highly abundant sea star *Asterina miniata* grazes the surface of rocky reefs and adjacent sand substrata, nonselectively foraging on benthic diatoms, microscopic stages of macroalgal, algal, and animal detritus, and sessile planktivores, such as barnacles and bryozoans (Pearse and Hines 1976, Leonard 1994). For most species, however, a review of the published literature identified clear primary trophic levels for consumers (Graham 2004), and assignment of species to these primary feeding-categories, as we do in the following sections, helps to identify key differences in trophic interactions within and between southern and central Californian kelp forests.

As with producers, there are conspicuous differences in the species composition of kelp forest consumers between central and southern California, mostly due to biogeographic differences in the major taxa. For example, reef fishes of tropically-derived families, including wrasses, sea basses, damselfishes, and gobies, are all more abundant and diverse south of Point Conception, whereas fishes of temperate-derived families, including rockfishes, sculpins, greenlings, and surfperches, are more abundant and diverse to the north (Miller and Lea 1972, Horn and Allen 1978, Stephens et al. 2005). In fact, the composition of consumer species can often differ greatly within regions in California, especially between the Californian mainland and Channel Islands, and among the Channel Islands (Ebeling et. al. 1980, and Patton et al. 1985 for fishes).

Planktivores and detritivores In both southern and central Californian kelp forests, the combined contributions of the detrital dissolved and particulate organics produced by *Macrocystis pyrifera,* and the constant influx of phyto- and zooplankton (for example mysids; Clarke 1971, Hobson and Chess 1976, Coyer 1985), support a striking abundance and diversity of filter and suspension feeding detritivores and planktivores (Harrold et al. 1988, Graham 2004). Particularly impressive among the detritivores and planktivores is the diversity of phyla, each of which in turn is represented by both high taxonomic diversity and abundance and, most notably, includes tunicates, sponges and anemones, bryozoans, gastropod and bivalve molluscs, annelids, echinoderms, crustaceans, and fishes. Some taxa are largely detritivorous, like the conspicuous and habitat-forming mounds of the colonial annelid *Diopatra ornata*, and some commercially sought sea cucumbers. Primarily planktivorous species are distributed throughout the water column, from the reef base to ocean surface. Those that position themselves well above the reef surface include a variety of fish species suspended in the water column (Hobson and Chess 1976) or sessile invertebrates attached to or closely associated with the surface of kelps and rocky promontories, such as mysids, caprellid amphipods, barnacles, and bryozoans (Wing and Clendenning 1971, Woolacott and North 1971, Bernstein and Jung 1979).

The most conspicuous differences among the planktivores in southern and central Californian kelp forests are among the fishes. Notably abundant in southern California are gobies, halfmoon, blacksmith, senorita, and several transient planktivorous species, such as *Salema*; all of which are members of tropically-derived families (Gobiidae, Kyphosidae, Pomacentridae, Labridae; Quast 1968a,b,c; Hobson and Chess 1976; Holbrook et al. 1990). In contrast, juveniles and adults of the temperate-derived rockfishes in the genus *Sebastes* are prevalent in central California (Miller and Geibel 1973, Hallacher and Roberts 1985, Singer 1985, Gaines and Roughgarden 1987, Carr 1991).

One notable difference in benthic planktivores between southern and central Californian rocky reefs is the prevalence of gorgonians in southern California, including *Lophogorgia chilensis, Muricea californica*, and *Eugorgia rubens* (Gotshall 1994), again reflecting the influence of tropical taxa in this region. However, most species employ their suspension and filter-feeding mechanisms to consume both detritus and plankton. The effects of these species on kelp forest dynamics can be substantial and are not limited to trophic attributes of the community. For example, in southern Californian forests, dense aggregations of the reef-building bivalve, *Chama arcana*, and the vermetid gastropod, *Serpulorbis squamigerus*, enhance the structural heterogeneity of rocky reef surfaces. Similarly, gorgonians and the sea cucumber *Pacythyone rubra* can carpet reef surfaces to the exclusion of other sessile species, including habitat-forming macroalgae like giant kelp (Patton et al. 1995).

Grazers and detritivores The great productivity of erect frondose algae on shallow subtidal reefs also supports a high diversity of herbivorous grazers. However, most grazers, perhaps with the exception of a few herbivorous fishes, also utilize detritus, including a variety of echinoderms, including sea urchins, sea cucumbers, and sea stars (Harrold and Pearse 1987), gastropods (Watanabe 1984, Schmitt 1985), and crustaceans, such as isopods, amphipods, shrimps, hermit crabs, and spider crabs (Hines 1982, Coyer 1985). One clear example of differences in grazer composition between southern and central

Californian kelp forests is the greater diversity of sea urchins in the south. Three species, *Strongylocentrotus pupuratus, S. franciscanus*, and *Lytechinus anamesus*, occur in particularly high densities in southern Californian forests and *Centrostephanus coronatus* is found only south of Point Conception (Harrold and Pearse 1987). Similarly, several predominant gastropod taxa such as *Lithopoma, Norrisia,* and (historically) several abalone (*Haliotis*) species are in greater abundance in southern Californian forests (Tucker 1954, Keen and Coan 1974). The seastar, *Asterina miniata*, is a prominent and ubiquitous detritivore-grazer in kelp forests throughout southern and central California (Leonard 1994). Like the planktivorous fishes mentioned above, the most abundant herbivorous reef fishes, the halfmoon, *Medialuna californiensis,* and the opaleye, *Girella nigricans*, are members of tropically-derived families (Kyphosidae and Girellidae, respectively) and are far more abundant south of Point Conception.

Primary predators A suite of carnivorous fishes, gastropods, crustaceans, and sea stars exploits this multitude of grazers and planktivores. Small reef fishes, including gobies, blennies, and especially the kelp fishes (family Clinidae), forage on the many small herbivorous crustaceans, notably amphipods that are associated with macroalgae, as do larger reef fishes, such as surfperches, painted greenlings, and kelp greenlings. Likewise, a number of carnivorous neogastropods feed on barnacles and other sessile invertebrates. The predatory gastropods, especially of the family Muricidae, as well as whelks (e.g., *Kelletia*) and cowries (e.g., *Cyprea*) are diverse and abundant in southern Californian forests.

Particularly voracious and abundant predators are the many sea stars that, like their intertidal counterparts, are responsible for creating and maintaining a patchy distribution of the benthic sessile fauna and for producing rocky surfaces available for settlement and recruitment of sessile species (Duggins 1983, Harrold and Pearse 1987). Due to their high mobility, large sea stars, such as the sunflower star, *Pycnopodia helianthoides*, also influence the local abundance of mobile grazers, particularly gastropods (Watanabe 1984) and sea urchins (Duggins 1983, Harrold and Pearse 1987). Because of the greater diversity and abundance of sea stars with increasing latitude, their role in kelp forest food webs seems to increase in more northerly forests, a pattern amplified by the fact that sea star populations in southern California have also experienced episodic disease outbreaks (Tegner and Dayton 1987).

Octopi are also voracious gastropod predators in kelp forests throughout California, but especially so in southern California (Rosenthal et al. 1974, Pearse and Hines 1976, Ambrose 1986). Predation by the largest crustaceans in the system, cancer crabs and the California spiny lobster (*Panulirus interruptus*), rivals that of sea stars and octopi. Indeed, one of the most conspicuous differences in the regime of primary predators between southern and central kelp forests is that *P. interruptus*, which preys upon bivalves, gastropods, and sea urchins (Mitchell et al. 1969, Tegner and Levin 1983, Robles 1987), occurs only south of Point Conception, where it is abundant.

Secondary predators Fishes; birds, including cormorants, herons, and egrets; marine mammals, including seals, sea lions, and sea otters; and humans eat kelp-forest planktivores, detritivores, grazers, and primary predators. Kelp-forest secondary predators constitute three feeding groups: species that primarily consume (1) fishes, (2) invertebrates, or (3) both fishes and invertebrates. Of the species that feed primarily on fishes, both birds and marine mammals are substantial sources of predation in both southern and central Californian forests (Foster and Schiel 1985). Whether or how the magnitude of predation by these groups varies regionally is not clear, especially because of the great spatial variation in their distribution and foraging activities within each region. For most kelp-forest fishes, however, the major source of mortality is predation by piscivorous fishes, reflected by both the massive mortality of the small juvenile stages that are vulnerable to predation by larger fishes and the predominance of small fishes in the diet of many piscivorous fishes (Steele and Anderson 2005, Carr and Syms 2006).

Given the differences in the representation of tropical and temperate families in central and southern kelp forests, it is not surprising that species composition of piscivores varies regionally. For example, the piscivorous sea basses, especially the kelp bass, are prevalent predators in the south (Limbaugh 1955, Young 1963, Quast 1968a) and the rockfishes are the most prevalent piscivores in the north (Hallacher and Roberts 1985). There are also conspicuous regional differences in species composition for invertebrate predators. Sea otters (*Enhydra lutris*) were extirpated from southern California in the early 1800s and have yet to permanently recolonize regions south of Point Conception (Riedman and Estes 1988), but they appear to play a particularly important role in limiting the density of the dominant herbivores, such as sea urchins, in central Californian forests (Harrold and Pearse 1987, Ebeling and Laur 1988, Estes and Harrold 1988). In southern Californian forests, their role is replaced in part by one fish, the California sheephead (Cowen 1983, 1986), and possibly a crustacean, the California spiny lobster, *Panulirus interruptus* (Tegner and Levin 1983, Lafferty 2004). These regional taxonomic differences, combined with regional differences in the relative productivity of both frondose algae and plankton, set the stage for marked differences in key species interactions that both establish and maintain the structure of kelp forests in central and southern California.

Food-web Structure

One property clearly common to southern and central Californian kelp forests is the fundamental importance of kelp, and primarily *Macrocystis pyrifera*, as an overwhelming source of primary production and detritus that fuels both the grazer-dependent and the detritus-dependent trophic pathways in these systems (fig. 5.3). The actual diversity of forest-dwelling species involved in either or both of these pathways has never been quantified, but clearly constitutes a major portion of the great diversity characteristic of these communities (Graham 2004). Thus, one consideration for increasing our understanding of the ecological processes that fundamentally structure Californian kelp forest communities is to determine how these systems change in response to variation in the presence or abundance of kelp.

It is obvious that localized kelp loss will have ecological consequences, since remaining nonkelp primary producers cannot replace the vast amounts of energy and habitat lost. But for any given system it is unknown whether the changes will come in terms of species biomass, which also provides habitat structure, productivity, or diversity. The inability of ecologists to predict such tangible consequences of kelp loss is due largely to our inability to isolate trophic from habitat linkages between kelp and their associated species, and to distinguish them from other species interactions or associations. For example, in Californian *Macrocystis* forests some generalist taxa, such as sea urchins, receive the bulk of their energy from giant kelp drift when it is present (Pearse and Hines 1976, Harrold and Reed 1985), suggesting a strong trophic linkage. Nonetheless, sea urchins can survive for years to decades in the almost complete absence of large attached macroalgae (Harrold and Pearse 1987); the abundance of abalones, however, appears to be more dependent on a constant source of drift kelp as abalone are lost from some southern Californian systems following episodic deforestation (Graham 2004).

Habitat linkages may be just as problematic since some taxa, such as kelp surf perch (*Brachyistius frenatus*) and the kelp goby (*Lethops connectens*), appear to have essentially obligate associations with giant kelp structure at some stage in their life history (Graham 2004), whereas others, such as Norris's top snail (*Norrisia norrisi*), giant kelp fish (*Heterostichus rostratus*), kelp bass (*Paralabrax clathratus*), kelp rockfish (*Sebastes atrovirens*), and juveniles of several species of rockfishes, have only facultative relationships with the forest structure. Facultative species are often associating with various algae that are either present as an understory beneath sparse canopies or continuing to exist in the complete absence of kelp forests; yet all of these species decline with the removal of giant kelp (Miller and Geibel 1973; Bodkin 1988; Ebeling and Laur 1988; Carr 1989, 1991,

1994; DeMartini and Roberts 1990; Holbrook et al. 1990; Anderson 1994; Graham 2004; Stephens et al. 2005). The job of disentangling trophic from habitat associations seems daunting given that many taxa have both trophic and habitat linkages, omnivory is poorly understood, and the importance of indirect interactions in a community with >200 common species is unknown (Graham 2004).

No studies have addressed variability in the strength of trophic links in kelp forests across any temporal or spatial scale, although Sala and Graham (2002) did study among-species variability in interaction strength between *Macrocystis pyrifera* and its invertebrate grazers. Still, the last 50 years of ecological and natural history studies suggest that differences in the strength of trophic links exist between southern and central Californian forests. Some planktivores are noticeably more abundant in southern than in central Californian forests, including the blacksmith, a very abundant planktivorous damselfish (Hobson and Chess 1976, Ebeling et al. 1980, Bray 1981). Large standing biomasses of several species of planktivorous juvenile and adult rockfishes are, however, supported by great concentrations of zooplankton in the coastal waters of central California (Hallacher and Roberts 1985, Singer 1985, Gaines and Roughgarden 1987, Carr 1991). Many species of fish have juveniles that are numerous, planktivorous, and co-occur in both regions, including senoritas and kelp perch. High biomasses of planktivores in central Californian forests may reflect the greater coastal productivity in the upwelling-dominated ocean climate that characterizes inshore waters north of Point Conception (Strub and James 2000).

A second interesting property of Californian kelp forest food webs is that, despite the relatively high species richness of the kelp forest communities, few primary consumer taxa, such as sea urchins or amphipods, have been shown to overexploit producer populations. Furthermore, although abiotic factors can be key in controlling grazer outbreaks (Ebeling at al. 1985, Foster and Schiel 1988), when predation is found to be important, only a small set of predators is apparently necessary to control these grazer populations. Sea otters are voracious predators of sea urchins, but in California they are restricted primarily to kelp forests north of Point Conception (Riedman and Estes 1988). The effect of sea otters in central California may be reinforced by the greater abundance of two other sea-urchin predators that become increasingly abundant from south to north: the predatory sea star, *Pycnopodia helianthoides* (Duggins 1983), and the wolf eel, *Anarrichthyes occellatus* (Hulberg and Graber 1980). In the absence of this suite of predators in southern California, the spiny lobster (*Panulirus interruptus*) and the California sheephead (*Semicossyphus pulcher*) have been identified as potential sea-urchin predators. Large sea urchins may, however, have a size-refuge (Cowen 1983, Tegner and Levin 1983, Cowen 1986, Lafferty 2004).

Changes in sea urchin densities in response to either experimentally altered sheephead densities (Cowen 1983) or differences in lobster densities inside and outside marine reserves (Lafferty 2004) are consistent with the influential role of these two species in forests along the Californian Channel Islands. The southern Californian kelp forest has maintained a sea urchin fishery due to lower levels of natural predation in the absence of the sea otter. Finally, another key regional difference between these systems is that fisheries in southern California heavily target upper-level predators, such as lobster and sheephead, while sea otters are protected in central California.

The potential for overgrazing is not only regulated by predation. The strength of interactions between grazers and giant kelp forests in southern and central California may also reflect differences in the balance between kelp productivity and herbivory between these regions. In central California, no observations of overgrazing of *Macrocystis pyrifera* by the highly abundant herbivorous crustaceans that inhabits and grazes giant kelp plants have been recorded. In contrast, observations of herbivory by the amphipod, *Amphithoe humeralis*, leading to massive kelp biomass loss, have been recorded in southern Californian forests (Tegner and Dayton 1987, Dayton et al. 1998, Graham 2002). The ability of these small and abundant grazers to cause marked declines in forest biomass may reflect (1) a lower productivity potential of southern Californian kelps; (2) greater vulnerability of kelps to oceanographic variability in southern California, such as ENSO; or (3) greater effects of such oceanographic variability in southern California on the abundance of primary predators that may control grazers, specifically negative effects of ENSO on planktivorous fishes that control amphipod outbreaks (Tegner and Dayton 1987, Dayton et al. 1998, Graham 2002). Thus, the strength and effect of trophic interactions may vary regionally in response to large-scale variation in oceanographic processes that influence either producer or consumer populations, or both.

Temporal and Within-region Spatial Variation

Food Web Dynamics and Productivity

Southern and central Californian kelp forests share one fundamental trait: the reliance on large kelps as the primary source of energy and biogenic habitat. Still, large regional differences exist in species composition and the identity of key species, both of which appear to be due in a large part to broad-scale biogeographical processes. Within regions, however, the dynamics and productivity of kelp populations can also be highly variable in space and time. For example, the dynamics of perennial *Macrocystis pyrifera* populations on either side of the Monterey Peninsula (Hopkins Marine Station vs. Stillwater Cove) can be out of phase despite being separated by only 15 km of coastline; this pattern appears to be due to both variable exposure to ocean swell (northwest swell versus southwest swells, respectively) and the rock type of the reefs (see similar arguments by Foster and Schiel 1985, 1988). Four km south of the Monterey Peninsula along the central Californian coast, energy input to kelp forest systems is likely constrained by the restricted productivity of the annual kelp *Nereocystis luetkeana*, which can dominate the exposed open coastline (Foster and Schiel 1985). Finally, there appears to be little coherence to the dynamics of local *Macrocystis pyrifera* populations south of Point Conception, even those at opposite ends of the same kelp forest (Dayton et al. 1992, 1999), except during catastrophic warm ENSO events when most southern Californian kelp forests are simultaneously decimated (Edwards 2004).

Nearshore oceanographic processes can be important in regulating various aspects of the population biology of kelp forest species, including dispersal, nutrition, reproductive output, and survival, and thus environmental factors can directly cause fluctuations in kelp forest producer and consumer populations. Indeed, each kelp forest species must be considered to have a component of its population variance due to processes unrelated to associations with kelps. Well documented within-region variability in kelp population dynamics (Dayton et al. 1992, 1999) and known trophic and habitat associations among kelp forest species suggest, however, that kelp-associated processes may be responsible for much of the food-web dynamics over short spatial scales and a broad range of temporal scales. Furthermore, kelp-forest species may have trophic and habitat associations with particular kelp taxa, such as canopy-fishes obligatorily associated with *Macrocystis pyrifera* (Graham 2004). Although the extent of associations with particular kelp taxa have yet to be explored, geological and oceanographic processes that regulate the distribution and abundance of specific kelp taxa may also have important effects on the structure and dynamics of kelp forest food webs. In the next section, we discuss various methodological and theoretical challenges of disentangling the roles of trophic interactions from habitat associations in structuring kelp forest communities.

Ecological Consequences of Kelp Loss

Despite known trophic and habitat associations among kelp forest taxa, the inherent complexity of these communities makes it difficult to study community responses to continuous variability in kelp distribution and abundance. The approach of directly relating species abundance to kelp abundance (such as kelp frond density or biomass/m^2) has only proven feasible when working with small groups of closely-associated taxa (Holbrook et al. 1990). Consequently, community-based studies would be most beneficial if designed initially to focus attention on community responses to kelp loss, such as the presence or absence of kelp at a variety of temporal and spatial scales (Graham 2004), and some researchers have directly studied the effects of kelp loss on specific assemblages within kelp forest communities (Clark et al. 2004).

Experimental and large-scale long-term observational studies may be useful for studying the effects of kelp loss on community structure and food web dynamics. Specifically, experiments are useful for disentangling the many potentially confounding influences of processes that co-vary among and within forests over time, including kelp canopy biomass and detrital production, and to identify the causal relationships between components of the system. Monitoring studies, on the other hand, can reveal the strength of relationships between different community components across broad temporal and spatial scales, and help to define a broader context for interpreting conclusions drawn from more restricted experimental studies. Here, we introduce a series of experiments designed to disentangle trophic interactions from habitat associations, and argue for a broader role of large-scale, long-term monitoring studies of environmental correlates.

Experimental protocols One of the most exciting characteristics of kelp forest communities is that they are inherently amenable to experimental manipulations. Given their sessile habit, the distribution, abundance, and size of individual kelp plants and populations can be manipulated over ecologically relevant temporal scales of weeks to generations (Ambrose and Nelson 1982; Ebeling and Laur 1985; Bodkin 1988; Eckman et al. 1989; Carr 1989, 1994; Reed 1990; Dayton et al. 1999; Clark et al. 2004). Many questions await field experimentation: (1) what are the relative contributions of kelp production and habitat provision to kelp forest community structure? (2) To what extent does alteration of the physical environment by kelp—such as subsurface light, sedimentation, water motion, and turbulence—confound the provision of energy and habitat? (3) Is the positive effect of kelp presence on community function restricted to particular kelp taxa? (4) If so, what biological characteristics of these taxa are responsible for the effect? (5) What are the relative contributions of direct grazing versus phytodetritus production to consumer production? (6) How much variability in kelp forest dynamics and diversity can be accounted for by the effect of kelp on other energy and habitat providers, such as understory algae? (7) What are the relative contributions of phytoplankton versus frondose algal primary production?

Although broad in scope, each of these questions can be addressed either directly with manipulation of various components of local kelp stands, such as entire plants, canopy biomass, detritus, turf algae abundance, and kelp species composition, or in combination with ecological modelling (e.g., ECOPATH), trophic analyses (e.g., stoichiometry), and laboratory studies of species interactions (Anderson 2001). In theory, the design of such experiments is straight forward, and indeed each of these components has been manipulated to some extent by past researchers. The key for community-level studies, however, is to conduct the experiments over a broad enough range of spatial and temporal scales to ensure independence of treatment levels and relevance to entire kelp forest communities.

The mobility of many temperate species, especially fishes, and the slow growth and demographic responses of the many long-lived species necessitate large-scale and long-term press manipulations to assess the community-wide influence of kelp. Optimal experimental designs would include replicates of independent reefs with randomly allocated kelp treatments (no kelp versus kelp with canopy, versus kelp without canopy). The large size of most natural reefs along the Californian coast, however, makes this logistically problematic. Alternatively, large-scale removal of kelp (hundreds of meters) between replicates within a large continuous kelp forest could be used to create independent kelp treatments. In either case, water motion, import and export of phytodetritus, and immigration and emigration of consumers are likely to interact with kelp populations at scales >100 m (Jackson and Winant 1983, Jackson 1998), and thus efforts to expand the scale of such manipulations are critical to the advancement of our understanding of these systems. Furthermore, decoupling the relative effects of primary production and habitat structure on a forest's influence on community structure would entail artificial kelp structures akin to similar approaches used in seagrass systems (Bell et al. 1985), but at a more grand scale. Probably the most tractable experiments are those that directly manipulate kelp species composition, canopy structure and biomass, standing-stock of large phytodetritus or drift kelp, and grazer abundance. Because removal of kelp abundance simultaneously alters the energy source and the physical structure, decoupling of the two processes can only be achieved by holding one constant while removing the other. For example, with canopy removals, artificial kelp can be added to mimic physical structure and drift kelp can be added or removed to manipulate an energy source, although it is unknown how well manipulations of phytodetritus can be maintained in such fluid habitats. We consider that persistent removal of kelp canopy and phytodetritus, perhaps orthogonally to ascertain their relative and combined effects, may be feasible over reasonable spatial scales.

Role of long-term time series How do the structure and dynamics of kelp forest communities vary in relation to local and regional differences in geologic (substratum type and relief) and oceanographic (wave exposure and oceanic productivity) attributes? How does the inherent species richness of a region affect the role of kelps in the system? What is the role of episodic oceanographic events, such as ENSO, or regime shifts in regulating kelp–consumer associations? What are the relative contributions of nutrients from terrestrial and oceanic sources to kelp plant productivity and its utilization by kelp forest communities? The answers to such questions are vital to our understanding of the function of kelp systems, yet these processes vary over broad enough scales to be outside the range of field experimentation. As such, future studies will need to complement field experimentation with long-term monitoring of key kelp forest attributes such as kelp distribution and biomass, abundances of important consumer, environmental parameters, and to incorporate techniques not regularly utilized by kelp ecologists, including numerical modelling, genomics, stoichiometry, palaeontology, and archaeology.

In California, some organizations are collecting relevant ecological data on kelp forest community structure over broad temporal and spatial scales, for example the Channel Islands National Park kelp-forest monitoring program or the Partnership for Interdisciplinary Study of the Coastal Oceans (PISCO). These programs provide invaluable data to broaden the interpretation of small-scale field experiments that are replicated over broad spatial scales. Furthermore, numerous local, regional, state, and federal institutions and agencies are collecting long-term data sets of key parameters of particular species useful for long-term studies of kelp forest systems. For example, aerial photographs of kelp canopy area have been collected by various people and groups over many regions in California for the last 60 years, and have supported numerous ecological studies of kelp systems; such long-term monitoring of kelp canopies has been greatly enhanced by regular aerial surveys using digital photography and multi-spectral data collection. Hyperspectral surveys conducted by the Center for Integrative Coastal Observation Research and Education (CI-CORE) offers promise for remote sensing of specific kelp species, and even health and productivity of kelp canopies.

One new and exciting approach to understanding long-term change in kelp forest community structure is to explore the geologic and palaeontological records of human use of kelp forest resources to reconstruct the spatial chronology of forests within and between regions (Graham et al. 2003, Kinlan et al. 2005). In California, archaeological collections from human midden sites on the Channel Islands and southern and central Californian mainland include rich marine invertebrate and vertebrate assemblages extending back almost 13,000 years (Erlandson et al. 2005). In addition to providing simple estimates of the abundance of abalone, sea urchins, gastropods, bivalves, fishes, and mammals, many remains have a high level of organic matter preservation, which can lead to subsequent stable isotope and genetic analyses useful for the reconstruction of consumer diet and historical population size. Although such cross-disciplinary studies are rare for kelp systems, new technological advances, analytical tools, and a wealth of archived data provide numerous opportunities for cross-disciplinary explorations.

Human Influences

Aside from the regional extirpation of sea otters through the fur trade, anthropogenic disturbance to Californian kelp forest communities prior to 1900 came primarily from localized subsistence-level fishing. In the past century, however, and particularly in the past few decades, kelp communities have begun to experience dramatic increases in local and regional-scale pressure. These pressures include the entrainment of propagules, including spores, eggs, and larvae, by water intake systems, thermal pollution, increased turbidity, and sedimentation associated with cooling waters of coastal power plants; commercial and recreational fishing; and regional- to global-scale pressure from climate change. These shifts in the nature and the scale of human disturbance to kelp forest communities require new approaches for management and conservation. Some of these approaches are being implemented and are showing positive effects, such as establishment of marine protected areas, while other threats remain poorly addressed. Our focus here is to briefly discuss the potential causes of anthropogenic modification of kelp forest systems and its consequences for kelp populations.

Exploitation and Habitat Loss

The greatest direct impact to Californian kelp forest communities has come from human exploitation of mammals, fish, invertebrates, and kelp that make up these communities. Although this exploitation has been occurring for centuries, the rate and magnitude have increased significantly in the past few decades with the advent of new fisheries and new harvest methods. Sheephead (*Semicossyphus pulcher*), lobsters (*Panulirus interruptus*), abalone (*Haliotis* spp.), and many rockfishes (*Sebastes* spp.) have all been over-fished, some to ecological extinction (Dayton et al. 1998, Lafferty 2004), and in their place new fisheries on other rockfishes, sea cucumbers (*Parastichopus* spp.), sea urchins (*Strongylocentrotus* spp.), and sportfish like kelp bass (*Paralabrax clathratus*) have developed and are increasing.

These fisheries extract huge amounts of consumer biomass from kelp forest ecosystems annually (Leet et al. 2001). Although not exclusively from kelp forests, commercial landings in California over the last decade (1994–2003) have averaged 187,870 metric tons (data from PacFIN), even with large spatial and temporal closures to groundfish, rockfish, and other species during this time period. Recreational fisheries have removed an average of 10,000 fish per year from Californian waters over this same time period, nearly a quarter of which were rockfishes (data from RecFIN). These recreational fisheries catches are much smaller than the commercial catches, but often target different species, particularly those that reside in kelp forests, and may thus affect kelp forest ecosystems disproportionately.

Many of these fished species are long-lived and slow-growing abalone and rockfishes, and may not be able to recover quickly from heavy fishing pressure. Past changes in fishing regulations for some of these species have had mixed results in maintaining sustainable population levels, leading to complete closure of the fishery in some cases. The great reduction in these kelp forest species is not trivial (Dayton et al. 1998); the structure of these biological communities has fundamentally changed. Evidence suggests that the dramatic reduction in the population sizes of species like lobsters, sheephead, and sea otters may have driven some kelp forests to become sea urchin barrens (Tegner and Dayton 2000, Behrens and Lafferty 2004, Lafferty 2004). Because of the relatively slow growth of many sea urchin predators, the reversal of such community state shifts may take a decade or more even if fishing were to cease completely (Shears and Babcock 2003). Still, despite most of the southern Californian kelp forests lacking sizable populations of these predators, the forests have not collapsed (Steneck et al. 2002); Foster and Schiel (1988) alternatively suggest that physical processes affecting sea urchin mortality and recruitment, rather than predation, control sea urchin population explosions in California.

Giant kelp, *Macrocystis pyrifera*, has also been harvested for decades for use in a wide variety of food, cosmetic, and fertilizer products. Although this harvest appears to have little if any effect on the kelp (Kimura and Foster 1984) (since kelps have extremely rapid growth rates [North 1994]), not much is known about how kelp harvesting may affect the species that use kelp canopy as habitat. In particular, many rockfish (*Sebastes* spp.), kelp bass (*Paralabrax clathratus*), kelp surfperch (*Brachyistius frenatus*), and other species of fish are known to use the canopy as a nursery habitat. A few studies have examined the effects of kelp harvesting on associated fish populations or, in particular, the removal of kelp canopy as nursery habitat (Limbaugh 1955, Davies 1968, Quast 1968d, Miller and

Geibel 1973). Most of these studies have been conducted in southern California and all have been limited in spatial scale or replication. It remains unknown how the timing or extent of kelp harvest influences post-settlement survival or the population dynamics of most fish species.

Coastal pollution (Swartz et al. 1983), power plant operations (Schroeter et al. 1993, Reitzel et al. 1994, Bence et al. 1996, Schiel et al. 2004), and the dredging of channels and harbors (North and Schaefer 1964) have negatively affected kelp forests. Although smaller in scale than the fishing effects described above, habitat loss and pollution have dramatically reduced the size of local kelp forests and the abundances of the associated biological communities (North 1971, Dayton et al. 1984, Schroeter et al. 1993, Reitzel et al. 1994, Bence et al. 1996, Schiel et al. 2004), the effect of episodic oil spills on Californian kelp systems has been relatively minor (Foster et al. 1971). As humans continue to migrate to coastal areas, this pressure will certainly increase, although the effects can sometimes be counterintuitive. For example, the large sewage spill that occurred in the Point Loma kelp forest actually benefited kelp populations by stimulating recruitment through nutrient input (Tegner et al. 1995).

Invasive Species

The spread of nonnative species into habitats and locations continues to be a global problem, yet the potential effect of invasive species on kelp forest systems is largely unknown. For example, *Sargassum muticum* has spread throughout the Northeast Pacific and persisted for many decades (Druehl 1973, Norton 1981). Its ability to rapidly colonize and cover completely canopy-free areas can prevent the reestablishment of giant kelp forests (Ambrose and Nelson 1982), although these effects appear to be limited in time and space (Foster and Schiel 1992). Two species of invasive seaweeds have only recently been introduced to California: the siphoneous green alga *Caulerpa taxifolia* (Williams and Grosholz 2002) and the Asiatic kelp, *Undaria pinnatifida* (Silva et al. 2002, Thornber et al. 2004). In California, neither species has been reported on natural substratum along the open coast, but both species have been documented to alter benthic community structure in other regions of the world where they have become abundant (Piazzi et al. 2001, Valentine and Johnson 2003, Casas et al. 2004). In the cases where *Undaria* has had an effect on natural kelp populations in Tasmania and Argentina, the inherent richness of the local kelp assemblages are an order of magnitude lower than in California (Valentine and Johnson 2003, Casas et al. 2004). It remains to be seen whether the increased diversity of Californian kelp forests can buffer them from the ecological and economic threats of invasion.

Climate Change

Changes in global and regional climate regimes are expected to affect Californian kelp-forest communities. Kelps have limited depth and temperature ranges; as sea level and surface temperature (SST) rise with global warming, kelp distributions will be modified according to subsequent changes in distribution of the substratum (rocky reefs) and productivity regimes amenable for kelp attachment and growth. For example, Holocene sea-level rises likely led to large changes in total area of kelp forest habitat around the Californian Channel Islands as broad nearshore rocky platforms shrank (Graham et al. 2003, Kinlan et al. 2005). This shift coincided with conspicuous changes in total biomass of kelp-associated species, such as abalone, sea urchins, and turban snails in Native American shell middens on the Channel Islands (Erlandson et al. 2005). Similarly, past annual and decadal shifts in regional oceanographic temperature regimes have shifted the southern range limit of kelp in Baja California, Mexico over 100 km to the north (Hernandez-Carmona et al. 1989). If global climate change continues to drive SST higher, the southern limit of kelp distributions is expected to move farther north along the Baja and southern Californian coasts, depending on the magnitude of change in SST.

Finally, it has been suggested that climate change may be increasing the frequency of ENSO events (Diaz et al. 2001), which can have deleterious effects on kelp forests due to short-term increases in SST and the intensity and frequency of storms and decreases in nutrient concentrations (Dayton and Tegner 1984, Dayton et al. 1999, Edwards 2004). The combined pressures on kelps of higher SST and disturbance frequency in the southern end of the range may drive kelp range limits farther north than would be predicted from either factor alone. Recent models of regional-scale effects of climatic change suggest that changes in the temperature differential between land and sea will alter coastal wind fields leading to changes in the frequency, magnitude, and location of coastal upwelling that fuels kelp productivity (Bakun 1990, Diffenbaugh et al. 2004).

Management and Intervention

Efforts to control and manage human effects on kelp forest communities have for the most part focused on moving sewage discharges offshore and managing fishing effort. Additionally, huge spatial and complete fisheries closures have recently been implemented in attempts to restore depleted groundfish, such as the rockfish fishery. In response to the potentially unsustainable way that nearshore fish stocks have traditionally been exploited, recent efforts have turned to marine protected areas (MPAs) including no-take marine reserves as tools to complement management and protect multiple species simultaneously. Marine reserves have been shown to increase population sizes and species diversity within reserves in kelp forests around the world (Edgar and Barrett 1999, Babcock et al. 1999, Halpern 2003), although the reserves are most beneficial to target species and the effects on populations outside reserve boundaries will vary depending on patterns of larval dispersal (Gaines et al. 2003, Shanks et al. 2003, Palumbi 2003) and species mobility (Chapman and Kramer 2000), among other factors. Reserves incorporating kelp forests may also provide protection from kelp harvest to species using the canopy as nursery habitat, although the nursery contribution of kelp to growth and survival of juveniles relative to other habitats has yet to be clearly documented or quantified.

Although reserves are likely to be able to provide protection from extraction of entire suites of species at once, they will be unable to account for threats to kelp forest communities caused by climate change. In fact, it may be difficult if not impossible to stop climate-driven changes from occurring, and so management efforts will have to be designed to account for rather than protect from these changes. Unfortunately, few if any current efforts to manage or protect kelp forest communities are accounting for the potential effects of climate change on these communities. Long-term protection is likely to be most successful if marine reserves are placed and sized to account for shifting species ranges (both across latitudes and depths) in response to global climate change.

References

Abbott, I.A., and G.J. Hollenberg. 1976. Marine algae of California. Stanford University Press, Palo Alto, CA.

Ambrose, R.F. 1986. Effects of octopus predation on motile invertebrates in a rocky subtidal community. Marine Ecology Progress Series **30**:261–273.

Ambrose, R.F., and B.V. Nelson. 1982. Inhibition of giant kelp recruitment by an introduced alga. Botanica Marina **25**:265–267.

Anderson, T.W. 1994. Role of macroalgal structure in the distribution and abundance of a temperate reef fish. Marine Ecology Progress Series **113**:279–290.

Anderson, T.W. 2001. Predator responses, prey refuges, and density-dependent mortality of a marine fish. Ecology **82**:245–257.

Austin, W.C. 1985. An annotated checklist of marine invertebrates of the cold temperate Northeast Pacific. Khoyatan Marine Laboratory, Cowichan Bay, British Columbia, Canada.

Babcock, R.C., S. Kelly, N.T. Shears, J.W. Walker, and T.J. Willis. 1999. Changes in community structure in temperate marine reserves. Marine Ecology Progress Series **189**:125–134.

Bakun, A. 1990. Global climate change and intensification of coastal ocean upwelling. Science **247**:198–201.

Batchelder, H.P., and T.M. Powell. 2002. Physical and biological condition and processes in the northeast Pacific Ocean. Progress in Oceanography **53**:105–114.

Behrens, M., and K. Lafferty. 2004. Effects of marine reserves and urchin disease on southern Californian rocky reef communities. Marine Ecology Progress Series **279**:129–139.

Bell, J.D., A.S. Steffe, and M. Westoby. 1985. Artificial seagrass: how useful is it for field experiments on fish and macroinvertebrates? Journal of Experimental Marine Biology and Ecology **90**:171–177.

Bence, J.R., A. Stewart-Oaten, and S.C. Schroeter. 1996. Estimating the size of an effect from a Before-After-Control-Impact paired series design: the predictive approach applied to a power plant study. Pages 133–149 *in* Schmitt, R.J., and C.W. Osenberg, editors. Detecting ecological impacts: concepts and applications in coastal habitats. Academic Press, San Diego.

Bernstein, B.B., and N. Jung. 1979. Selective processes and co-evolution in a kelp canopy community in southern California. Ecological Monographs **49**:335–355.

Bodkin, J.L. 1988. Effects of kelp removal on associated fish assemblages in central California. Journal of Experimental Marine Biology and Ecology **117**:227–238.

Bograd, S.J., D.A. Checkley and W.S. Wooster, editors. 2003. CalCOFI: a half-century of physical, chemical and biological research in the California current system. Deep Sea Research Part II: Topical Studies in Oceanography **50**(14–16):2349–2594.

Bray, R.N. 1981. Influence of water currents and zooplankton densities on daily foraging movements of blacksmith, *Chromis punctipinnis*, a planktivorous reef fish. Fishery Bulletin **78**:829–841.

Bray, R.N., and A.W. Ebeling. 1975. Food, activity, and habitat of three "picker-type" microcarnivorous fishes in the kelp forests off Santa Barbara, California. Fishery Bulletin **73**:815–829.

Breda, V.A., and M.S. Foster. 1985. Composition, abundance, and phenology of foliose red algae associated with two central California kelp forests. Journal of Experimental Marine Biology and Ecology **94**:115–130.

Briggs, K.T., W.B. Tyler, D.B. Lewis, and D.R. Carlson. 1987. Bird communities at sea off California: 1975 to 1983. Cooper Ornithological Society, Columbus, Ohio.

Carr, M.H. 1989. Effects of macroalgal assemblages on the recruitment of temperate zone reef fishes. Journal of Experimental Marine Biology and Ecology **126**:59–76.

Carr, M.H. 1991. Habitat selection and recruitment of an assemblage of temperate zone reef fishes. Journal of Experimental Marine Biology and Ecology **146**:113–137.

Carr, M.H. 1994. Effects of macroalgal dynamics on recruitment of a temperate reef fish. Ecology **75**:1320–1333.

Carr, M.H., and C. Syms. 2006. Recruitment. Pages 411–427 *in* Allen, L.G., D.J. Pondella II, and M.H. Horn, editors, The ecology of California marine fishes. University of California Press, Berkeley, Calif.

Casas, G., R. Scrosati, and M.L. Piriz. 2004. The invasive kelp *Undaria pinnatifida* (Phaeophyceae, Laminariales) reduces native seaweed diversity in Nuevo Gulf (Patagonia, Argentina). Biological Invasions **6**:411–416.

Chapman, M.R., and D.L. Kramer. 2000. Movements of fishes within and among fringing coral reefs in Barbados. Environmental Biology of Fishes **57**:11–24.

Chelton, D.B., and R.E. Davis. 1982. Monthly mean sea-level variability along the west coast of North America. Journal of Physical Oceanography **12**:757–784.

Clark, R.P., M.S. Edwards, and M.S. Foster. 2004. Effects of shade from multiple kelp canopies on an understory algal assemblage. Marine Ecology Progress Series **267**:107–119.

Clarke, W.D. 1971. Mysids of the southern kelp region. Nova Hedwigia **32**:369–380.

Cowen, R.K. 1983. The effect of sheephead (*Semicossyphus pulcher*) predation on red sea urchin (*Strongylocentrotus franciscanus*) populations: an experimental analysis. Oecologia **58**:249–255.

Cowen, R.K. 1986. Site-specific differences in the feeding ecology of the California sheephead, *Semicossyphus pulcher* (Labridae). Environmental Biology of Fishes **16**:193–203.

Coyer, J.A. 1985. The invertebrate assemblage associated with the giant kelp *Macrocystis pyrifera* at Santa Catalina Island, California, USA: a general description with emphasis on amphipods, copepods, mysids and shrimps. Fishery Bulletin **82**:55–66.

Coyer, J.A. 1987. The mollusk assemblage associated with the fronds of giant kelp *Macrocystis pyrifera* at Santa Catalina Island, California, USA. Bulletin of the Southern California Academy of Sciences **85**:129–138.

Darwin, C.R. 1839. Journal of researches into the geology and natural history of the various countries visited by the H.M.S. Beagle. Henry Colburn, England.

Davies, D.H. 1968. Statistical analysis of the relation between kelp harvesting and sportfishing in the California kelp beds. California Department of Fish and Game **139**:151–212.

Dawson, M.N. 2001. Phylogeography in coastal marine animals: a solution from California? Journal of Biogeography **28**:723–736.

Dayton, P.K., and M.J. Tegner. 1984. Catastrophic storms, El Nino, and patch stability in a southern California kelp community. Science **224**:283–285.

Dayton, P.K., V. Currie, T. Gerrodette, B.D. Keller, R. Rosenthal, and D. Ven Tresca. 1984. Patch dynamics and stability of some California kelp communities. Ecological Monographs **54**:253–289.

Dayton, P.K, Tegner M.J., Parnell P.E., and Edwards P.B. 1992. Temporal and spatial patterns of disturbance and recovery in a kelp forest community. Ecological Monographs **62**:421–445.

Dayton, P.K., M.J. Tegner, P.B. Edwards, and K.L. Riser. 1998. Sliding baselines, ghosts, and reduced expectations in kelp forest communities. Ecological Applications **8**:309–322.

Dayton, P.K., M.J. Tegner, P.B. Edwards, and K.L. Riser. 1999. Temporal and spatial scales of kelp demography: the role of oceanographic climate. Ecological Monographs **69**:219–250.

DeMartini, E.E., and D.A. Roberts. 1990. Effects of giant kelp (*Macrocystis*) on the density and abundance of fishes in a cobble-bottom kelp forest. Bulletin of Marine Science **46**:287–300.

Diaz, H.F., M.P. Hoerling, and J.K. Eischeid. 2001. ENSO variability, teleconnections and climate change. International Journal of Climatology **21**:1845–1862.

Diffenbaugh, N.S., M.A. Snyder, and L.C. Sloan. 2004. Could CO_2-induced land-cover feedbacks alter near-shore upwelling regimes? Proceedings of the National Academy of Sciences **101**:27–32.

Druehl, L.D. 1970. The patterns of Laminariales distribution in the northeast Pacific. Phycologia **9**:237–247.

Druehl, L.D. 1973. Marine transplantation. Science **179**:12.

Duggins, D.O. 1983. Starfish predation and the creation of mosaic patterns in a kelp-dominated community. Ecology **64**:1610–1619.

Duggins, D.O. 1988. The effects of kelp forests on nearshore environments: Biomass, detritus, and altered flow. Pages 192–201 *in* G.R.VanBlaricom and J.A. Estes, editors. The community ecology of sea otters. Springer-Verlag, Heidelberg, Germany.

Duggins, D.O., C.A. Simenstad, and J.A. Estes. 1989. Magnification of secondary kelp detritus in coastal marine ecosystems. Science **245**:170–173.

Ebeling, A.W., R.J. Larson, and W.S. Alevizon. 1980. Habitat groups and island-mainland distribution of kelp-bed fishes off Santa Barbara, California. Pages 403–432 *in* D.M. Power, editor, Multidisciplinary symposium on the California Islands. Santa Barbara Museum of Natural History, Santa Barbara, Calif.

Ebeling, A.W., and D.R. Laur. 1985. The influence of plant cover on surfperch abundance at an offshore temperate reef. Environmental Biology of Fishes **12**:169–179.

Ebeling, A.W., and D.R. Laur. 1986. Foraging in surfperches: resource partitioning or individualistic response? Environmental Biology of Fishes **16**:123–133.

Ebeling, A.W., and D.R. Laur. 1988. Fish populations in kelp forests without sea otters: effects of severe storm damage and destructive sea urchin grazing. Pages 169–191 *in* G.R. VanBlaricom and J.A. Estes, editors, The community ecology of sea otters. Springer-Verlag, Heidelberg, Germany.

Ebeling, A.W., D.R. Laur, and R.J. Rowley. 1985. Severe storm disturbances and reversal of community structure in a southern California kelp forest. Marine Biology **84**:287–294.

Eckman, J.E., D.O. Duggins, and A.T. Sewell. 1989. Ecology of understory kelp environments. I. Effects of kelp on flow and particle transport near the bottom. Journal of Experimental Marine Biology and Ecology **129**:173–187.

Edgar, G.J., and N.S. Barrett. 1999. Effects of the declaration of marine reserves on Tasmanian reef fishes, invertebrates, and plants. Journal of Experimental Marine Biology and Ecology **242**:107–144.

Edwards, M.S. 2004. Estimating scale-dependency in disturbance impacts: El Niños and giant kelp forests in the northeast Pacific. Oecologia **138**:436–447.

Erlandson, J.M., T.C. Rick, J.A. Estes, M.H. Graham, T.J. Braje, and R.L. Vellanoweth. 2005. Sea otters, shellfish, and humans: a 10,000-year record from San Miguel Island, California. Proceedings of the California Islands Symposium **6**:56–68.

Estes, J.A., and C. Harrold. 1988. Sea otters, sea urchins, and kelp beds: some questions of scale. Pages 116–150 *in* G.R. VanBlaricom and J.A. Estes, editors. The community ecology of sea otters. Springer-Verlag, Heidelberg, Germany.

Foster, M.S., M. Neushul, and R. Zingmark. 1971. The Santa Barbara oil spill. Part 2: Initial effects on intertidal and kelp bed organisms. Environmental Pollution **2**:115–134.

Foster, M.S., and D.R. Schiel. 1985. The ecology of giant kelp forests in California: a community profile. United States Fish and Wildlife Service Biological Report **85**:1–152.

Foster, M.S., and D.R. Schiel. 1988. Kelp communities and sea otters: keystone species or just another brick in the wall? Pages 92–115 *in* G.R. VanBlaricom and J.A. Estes, editors, The community ecology of sea otters. Springer-Verlag, Heidelberg, Germany.

Foster, M.S., and D.R. Schiel. 1992. Zonation, El Nino disturbance, and the dynamics of subtidal vegetation along a 30 m depth gradient in two giant kelp forests. Pages 151–162 *in* C.N. Battershill, et al., editors, Proceedings of the Second International Temperate Reef Symposium. NIWA Marine, Wellington, New Zealand.

Gaines, S.D., B. Gaylord, and J.L. Largier. 2003. Avoiding current oversights in marine reserve design. Ecological Applications **13**:S32-S46.

Gaines, S.D., and J. Roughgarden. 1987. Fish in offshore kelp forests affect recruitment to intertidal barnacle populations. Science **235**:479–481.

Gerard, V.A. 1976. Some aspects of material dynamics and energy flow in a kelp forest in Monterey Bay, California. Unpublished Ph.D. dissertation, University of California, Santa Cruz.

Gotshall, D.W. 1994. Guide to marine invertebrates: Alaska to Baja California. Sea Challengers, Monterey, Calif.

Graham, M.H. 2002. Prolonged reproductive consequences of short-term biomass loss in seaweeds. Marine Biology **140**:901–911.

Graham, M.H. 2004. Effects of local deforestation of the diversity and structure of southern California giant kelp forest food webs. Ecosystems **7**:341–357.

Graham, M.H., P.K. Dayton, and J.M. Erlandson. 2003. Ice ages and ecological transition on temperate coasts. Trends in Ecology and Evolution **18**:33–40.

Greene, H.G., and M.P. Kennedy, editors. 1986, 1987, 1989. Geologic map series of the California continental margin: California Division of Mines and Geology, Area 1 through 7, Maps 1A through 7D.

Hallacher, L.E., and D.A. Roberts. 1985. Differential utilization of space and food by the inshore rockfishes (Scorpaenidae: *Sebastes*) of Carmel Bay, California. Environmental Biology of Fishes **12**:91–110.

Halpern, B.S. 2003. The impact of marine reserves: do reserves work and does reserve size matter? Ecological Applications **13**:S117-S137.

Harrold, C., and J.S. Pearse. 1987. The ecological role of echinoderms in kelp forests. Pages 137–233 *in* M. Jangoux and J.M. Lawrence, editors,

Echinoderm studies, Volume 2. A. A. Balkema Press, Rotterdam, the Netherlands.

Harrold, C., and D.C. Reed. 1985. Food availability, sea urchin grazing, and kelp forest community structure. Ecology **66**:1160–1169.

Harrold, C., J. Watanabe, and S. Lisin. 1988. Spatial variation in the structure of kelp forest communities along a wave exposure gradient. Marine Ecology **9**:131–156.

Hernandez-Carmona, G., Y.E. Rodriguez-Montesinos, J.R. Torres-Villegas, I. Sanchez-Rodriguez, and M.A. Vilchis. 1989. Evaluation of *Macrocystis pyrifera* (Phaeophyta, Laminariales) kelp beds in Baja California, Mexico I Winter 1985–1986. Ciencias Marinas **15**:1–27.

Hickey, B.M. 1993. Physical oceanography. Pages 19–70 *in* M.D. Dailey, D.J. Reish, and J.W. Anderson, editors, Ecology of the Southern California Bight: a synthesis and interpretation. University of California, Berkeley, Calif.

Hickey, B.M. 1998. Coastal oceanography of Western North America from the tip of Baja California to Vancouver Island. Pages 349–393 *in* A.R. Robinson and K.H. Brink, editors, The sea, Volume 12. Wiley, N.Y.

Hines, A.H. 1982. Coexistence in a kelp forest: size, population dynamics, and resource partitioning in a guild of spider crabs (Brachyura, Majidae). Ecological Monographs **52**:179–198.

Hobson, E.S., and J.R. Chess. 1976. Trophic interactions among fishes and zooplankters near shore at Santa Catalina Island, California. Fishery Bulletin **74**:567–598.

Hobson, E.S., and J.R. Chess. 1986. Relationships among fishes and their prey in a nearshore sand community off southern California. Environmental Biology of Fishes **17**:201–226.

Holbrook, S.J., M. H. Carr, R.J. Schmitt, and J.A. Coyer. 1990. Effects of giant kelp on local abundance of reef fishes: the importance of ontogenetic resource requirements. Bulletin of Marine Science **47**:104–114.

Holbrook, S.J, Schmitt, R.J, and J.S. Stephens Jr. 1997. Changes in an assemblage of temperature reef fishes associated with a climate shift. Ecological Applications **7**:1299–1310.

Horn, M.H., and L.G. Allen. 1978. A distributional analysis of California coastal marine fishes. Journal of Biogeography **5**:23–42.

Hulberg, L.W., and P. Graber 1980. Diet and behavioral aspects of the wolf-eel, *Anarrhichthys ocellatus*, on sandy bottom in Monterey Bay, California. California Department of Fish and Game **66**:172–177.

Jackson, G.A. 1998. Currents in the high drag environment of a coastal kelp stand off California. Continental Shelf Research **17**:1913–1928.

Jackson, G.A., and C.D. Winant. 1983. Effect of a kelp forest on coastal currents. Continental Shelf Research **2**:75–80.

Keen, A.M., and E. Coan. 1974. Marine Molluscan genera of North America. Stanford University Press, Palo Alto, Calif.

Kimura, R.S., and M.S. Foster. 1984. The effects of harvesting *Macrocystis pyrifera* on the algal assemblage in a giant kelp forest. Hydrobiologia **117**:425–428.

Kinlan, B.P., M.H. Graham, and J.M. Erlandson. 2005. Late Quaternary changes in the size and shape of the California Channel Islands:

implications for marine subsidies to terrestrial communities. Proceedings of the California Islands Symposium **6**:119–130.

Lafferty, K.D. 2004. Fishing for lobsters indirectly increases epidemics in sea urchins. Ecological Applications **14**:1566–1573.

Leet W.S., C.M. Dewees, R. Klingbeil, and E.J. Johnson, editors. 2001. California's living marine resources: a status report. State of California Resources Agency and Fish and Game, Sacramento, Calif.

Legg, M.R. 1991. Developments in understanding the tectonic evolution of the California Continental Borderland. Pages 291–312 *in* Osborne, R.H., editor, From shoreline to abyss. Society for Sedimentary Geology, San Francisco, Calif.

Leonard, G.H. 1994. Effect of the bat star *Asterina miniata* (Brandt) on recruitment of the giant kelp *Macrocystis pyrifera* C Agardh. Journal of Experimental Marine Biology and Ecology **179**:81–98.

Limbaugh, C. 1955. Fish life in the kelp beds and the effects of kelp harvesting. University of California Institute of Marine Resources, IMR Reference 55–9. San Diego, Calif.

Lluch-Cota, D.B., W.S. Wooster, and S.R. Hare. 2001. Sea surface temperature variability in coastal areas of the Northeastern Pacific related to the El Nino-Southern Oscillation and the Pacific Decadal Oscillation. Geophysical Research Letters **28**:2029–2032.

Lynn, R.J., and J.J. Simpson. 1987. The California Current system: The seasonal variability of its physical characteristics. Journal of Geophysical Research **92**:12947–12966.

Miller, D.J., and J.J. Geibel. 1973. Summary of blue rockfish and lingcod life histories; a reef ecology study; and giant kelp, *Macrocystis pyrifera*, experiments in Monterey Bay, California. California Department of Fish and Game **158**:1–137.

Miller, D.J., and R.N. Lea. 1972. Guide to the coastal marine fishes of California. California Department of Fish and Game **157**:1–235.

Mitchell, C.T., C.H. Turner, and A.R. Strachan. 1969. Observations on the biology and behavior of the California spiny lobster, *Panulirus interruptus* (Randall). California Department of Fish and Game **55**:121–131.

Newman, W.A. 1979. California transition zone: significance of short-range endemics. Pages 399–416 *in* J. Gray and A J. Boucot, editors, Historical biogeography, plate tectonics and the changing environment. Oregon State University Press, Corvallis, Oreg.

North, W.J. 1971. The biology of giant kelp beds (*Macrocystis*) in California: introduction and background. Nova Hedwigia **32**:1–68.

North, W.J. 1994. Review of *Macrocystis* biology. Pages 447–527 *in* I. Akatsuka, editor, Biology of Economic Algae. Academic Publishing, the Hague, the Netherlands.

North, W.J., and M.B. Schaefer. 1964. An investigation of the effects of discharged wastes on kelp. The Resources Agency of California, State Water Quality Control Board, Pub. # 26, Sacramento, Calif.

Norton, T.A. 1981. *Sargassum muticum* on the Pacific coast of North America. Proceedings of the International Seaweed Symposium **8**:449–456.

Orr, R.T., and R.C. Helm. 1989. Marine mammals of California. University of California Press, Berkeley, Calif.

Palumbi, S.R. 2003. Population genetics, demographic connectivity, and the design of marine reserves. Ecological Applications **13**:S146–S158.

Patton, M.L., R.S. Grove, and R.F. Harman. 1985. What do natural reefs tell us about designing artificial reefs in southern California? Bulletin of Marine Science **37**:279–298.

Patton, M.L., R.S. Grove, and L.O. Honma. 1995. Substrate disturbance, competition from sea fans (*Muricea* sp.) and the design of an artificial reef for giant kelp (*Macrocystis*). Pages 47–59 *in* Proceedings from the International Conference on Ecological System Enhancement Technology for Aquatic Environments. Japan International Marine Science and Technology Federation, Tokyo, Japan.

Pearse, J.S., and A.H. Hines. 1976. Kelp forest ecology of the central California coast. Pages 8–9 in University of California, Sea Grant College Program annual report 1975–76. Sea Grant Publication 57, La Jolla, Calif.

Piazzi, L., G. Ceccherelli, and F. Cinelli. 2001. Threat to macroalgal diversity: effects of the introduced green alga *Caulerpa racemosa* in the Mediterranean. Marine Ecology Progress Series **210**:149–159.

Quast, J.C. 1968a. Observations on the food of the kelp bed fishes. Pages 109–142 in W.J. North and C.L. Hubbs, editors, Utilization of kelp-bed resources in southern California. State of California Resources Agency and Fish and Game, Sacramento, Calif.

Quast, J.C. 1968b. Fish fauna of the rocky inshore zone. Pages 35–55 *in* W.J. North and C.L. Hubbs, editors, Utilization of kelp-bed resources in southern California. State of California Resources Agency and Fish and Game, Sacramento, Calif.

Quast, J.C. 1968c. Estimates of the populations and standing crop of fishes. Pages 57–79 *in* W.J. North and C.L. Hubbs, editors, Utilization of kelp-bed resources in southern California. State of California Resources Agency and Fish and Game, Sacramento, Calif.

Quast, J.C. 1968d. Effects of kelp harvesting on the fishes of the kelp beds. Pages 143–212 *in* W.J. North and C.L. Hubbs, editors, Utilization of kelp-bed resources in southern California. State of California Resources Agency and Fish and Game, Sacramento, Calif.

Reed, D.C. 1990. An experimental evaluation of density dependence in a subtidal algal population. Ecology **71**:2286–2296.

Riedman, M.L., and J.A. Estes. 1988. A review of the history, distribution and foraging ecology of sea otters. Pages 4–21 *in* G.R. VanBlaricom, and J.A. Estes, editors, The community ecology of sea otters. Springer-Verlag, Heidelberg, Germany.

Reitzel, J., M.H.S. Elwany, and J.D. Callahan. 1994. Statistical analyses of the effects of a coastal power plant cooling system on underwater irradiance. Applied Ocean Research **16**:373–379.

Ricketts, E.F., J. Calvin, and J.W. Hedgpeth. 1968. *Between Pacific Tides*. Stanford University Press, Palo Alto, Calif.

Robles, C. 1987. Predator foraging characteristics and prey population structure on a sheltered shore. Ecology **68**:1502–1514.

Rosenthal, R.J., W.D. Clarke, and P.K. Dayton. 1974. Ecology and natural history of a stand of giant kelp, *Macrocystis pyrifera*, off Del Mar, California. Fishery Bulletin **72**:670–684.

Sala, E., and M.H. Graham. 2002. Community-wide distribution of predator-prey interaction strength in kelp forests. Proceedings of the National Academy of Sciences **99**:3678–3683.

Schiel, D.R., and M.S. Foster. 1986. The structure of subtidal algal stands in temperate waters. Oceanography and Marine Biology: An Annual Review **24**:265–307.

Schiel, D.R., J.R. Steinbeck, and M.S. Foster. 2004. Ten years of induced ocean warming causes comprehensive changes in marine benthic communities. Ecology **85**:1833–1839.

Schmitt, R.J. 1985. Competitive interactions of two mobile prey species in a patchy environment. Ecology **66**:950–958.

Schroeter, S.C., J.D. Dixon, J. Kastendiek, R.O. Smith, and J.R. Bence. 1993. Detecting the ecological effects of environmental impacts: a case study for kelp forest invertebrates. Ecological Applications **3**:331–350.

Shanks, A.L., B.A. Grantham, and M.H. Carr. 2003. Propagule dispersal distance and the size and spacing of marine reserves. Ecological Applications **13**:S159–S169.

Shears, N.T., and R.C. Babcock. 2003 Continuing trophic cascade effects after 25 years of no-take marine reserve protection. Marine Ecology Progress Series **246**:1–16.

Silva, P.C., R.A. Woodfield, A.N. Cohen, L.H. Harris, and J.H.R. Goddard. 2002. First report of the Asian kelp *Undaria pinnatifida* in the northeastern Pacific Ocean. Biological Invasions **4**:333–338.

Singer, M.M. 1985. Food habits of juvenile rockfishes (*Sebastes*) in a central California kelp forest. Fishery Bulletin **83**:531–541.

Smith, R.I., and J.T. Carlton, editors. 1975. Light's manual: intertidal invertebrates of the central California coast. University of California Press, Berkeley, Calif.

Spalding, H., M.S. Foster, and J.N. Heine. 2003. Composition, distribution, and abundance of deep-water (>30 m) macroalgae in central California. Journal of Phycology **39**:273–284.

Steele, M.A., and T.W. Anderson. 2005. The role of predation in the ecology of California's marine fishes. Pages 428–448 *in* Allen, L.G., D.J. Pondella II, and M.H. Horn, editors. The Ecology of California Marine Fishes. University of California Press, Berkeley, Calif.

Steneck, R.S., M.H. Graham, B.J. Bourque, D. Corbett, J.M. Erlandson, J.A. Estes, and M.J. Tegner. 2002. Kelp forest ecosystems: biodiversity, stability, resilience and future. Environmental Conservation **29**:436–459.

Stephens, J.S, R.J. Larson, and D.J. Pondella II Jr. 2005. Rocky reefs and kelp beds. Pages 227–252 *in* L.G. Allen, D.J. Pondella II, and M.H. Horn, editors, The ecology of California marine fishes. University of California Press, Berkeley, Calif.

Strub, P.T., and C. James. 2000. Altimeter-derived variability of surface velocities in the California Current system: 2. Seasonal circulation and eddy statistics. Deep Sea Research II **47**:831–870.

Swartz, R.C., F.A. Cole, D.W. Schultz, and W.A. Deben. 1983. Ecological changes in the Southern California bight near a large sewage outfall; benthic conditions in 1980 and 1983. Marine Ecology Progress Series **31**:1–13.

Tegner, M.J., P.K. Dayton, P.B. Edwards, K.L. Riser, D.B. Chadwick, T.A. Dean, and L. Deysher. 1995. Effects of a large sewage spill on a kelp forest community: catastrophe or disturbance? Marine Environmental Research **40**:181–224.

Tegner, M.J., and P.K. Dayton. 1987. El Niño effects on southern California kelp forest communities. Advances in Ecological Research **17**:243–279.

Tegner, M.J., and P.K. Dayton. 2000. Ecosystem effects of fishing in kelp forest communities. Ices Journal of Marine Science **57**:579–589.

Tegner, M.J., and L.A. Levin. 1983. Spiny lobsters and sea urchins: analysis of a predator-prey interaction. Journal of Experimental Marine Biology and Ecology **73**:125–150.

Thornber, C.S., B.P. Kinlan, M.H. Graham, and J.J. Stachowicz. 2004. Population ecology of the invasive kelp *Undaria pinnatifida* in California: environmental and biological controls on demography. Marine Ecology Progress Series **268**:69–80.

Tucker, A.R. 1954. American seashells. Van Nostrand, N.Y.

Valentine, J.P., and C.R. Johnson. 2003. Establishment of the introduced kelp *Undaria pinnatifida* in Tasmania depends on disturbance to native algal assemblages. Journal of Experimental Marine Biology and Ecology **295**:63–90.

Ware, D.M., and R.E. Thompson. 2000. Interannual to multidecadal time-scale climate variations in the northeast Pacific. Journal of Climate **13**:3209–3220.

Watanabe, J.M. 1984. The influence of recruitment, competition and benthic predation on spatial distributions of 3 species of kelp forest gastropods. Ecology **65**:920–936.

Williams, S.L., and E.D. Grosholz. 2002. Preliminary reports from the *Caulerpa taxifolia* invasion in Southern California. Marine Ecology Progress Series **233**:307–310.

Wing, B.L., and K.A. Clendenning. 1971. Kelp surfaces and associated invertebrates. Nova Hedwigia **32**:319–342.

Woolacott, R.M., and W.J. North. 1971. Bryozoans of California and northern Mexico kelp beds. Nova Hedwigia **32**:455–480.

Young, P.H. 1963. The kelp bass (*Paralabrax clathratus*) and its fishery, 1947–1958. California Department of Fish and Game **122**:1–67.

6

Biodiversity and Food-Web Structure of a Galápagos Shallow Rocky-Reef Ecosystem

Rodrigo H. Bustamante, Thomas A. Okey, and Stuart Banks

The global significance of Galápagos ecosystems is indicated by designation of the archipelago as a World Heritage Site (UNESCO 1978), a Man and Biosphere Reserve (Perry 1984), and a Ramsar Wetlands Site (Ramsar 2004). In addition, Ecuador—the nation-steward—has declared 97% of Galápagos land as national park and encircled approximately 138,000 km^2 of oceanic and coastal environments within a multiple-use marine reserve network (Bensted-Smith 2002).

Despite this protection, the nearshore marine ecosystems of the Galápagos have been modified considerably by sequential depletions of populations of marine organisms by fisheries—both legal and illegal (Ruttenberg 2001, Bustamante et al. 2002c), and by extreme El Niño events (Robinson and Del Pino 1985, Glynn 1988). In this chapter we describe the food web of a Galápagos rocky reef ecosystem and the recent history of its exploitation, and point out management and conservation implications. We achieve this by featuring a quantitatively explicit Ecopath food web model for a shallow rocky reef ecosystem at Floreana Island, Galápagos, which is representative of south-central Galápagos shelf reefs. This food web analysis includes results of dynamic simulations conducted to provide functional indices as well as an overview of structural aspects of the food web, including biomass, interaction strength, and keystone effects.

Understanding Galápagos marine species and the food webs that connect them requires an understanding of the environment. We thus begin with a description of the environmental and climatic setting followed by an overview of the resulting biotic richness and endemicity of this "crossroads" marine ecosystem. We frame our concluding discussion of management and policy solutions with a description of the historical serial depletion of Galápagos marine populations and a discussion of its ecological implications.

The Environment

Physical Setting

The Galápagos Archipelago is located in the equatorial eastern Pacific Ocean, about 1000 km west of Ecuador's mainland coastline between 01°40´N and 01°25´S and 89°15´W and 92°00´W (fig. 6.1). The archipelago consists of more than 130 large and small islands and islets with a total surface area of about 50,130 km^2 and a coastline length estimated at around 1,800 km (Snell et al. 1995, 1996). The islands are the tops of volcanoes 1–3 million years old that emerged from the relatively shallow Galápagos Platform (Christie et al. 1992, Geist 1996), which is surrounded by deeper water (1,000–4,000 m). The combination of the location of the archipelago at the confluence of warm and cold surface currents and its geomorphology causes upwelling of nutrient-rich waters and complex physical oceanographic regimes. Combined with the isolation of the Galápagos archipelago, this has led to the development of unique and diverse marine communities (Colinvaux 1972, Wellington 1984, James 1991).

Geological History

The Galápagos Islands are of volcanic origin formed by a magma plume at a geologic "hotspot" (Christie et al. 1992, Harpp et al. 2002). They are located on the northern edge of the Nazca crustal plate, which is bounded by the Cocos Plate to the north and the Pacific Plate to the west. The hotspot is located at a fracture zone on the eastern side of the East Pacific Rise that divided the Cocos Plate from the Nazca Plate some 25–30 million years ago (mya).

The Nazca plate is moving east over the stationary hotspot, which extrudes new seamounts and islands such that the younger large islands of Isabela and Fernandina form the western parts of the archipelago and the older islands lie to the east. Recent estimates suggest that the oldest islands (San Cristóbal and Española) were

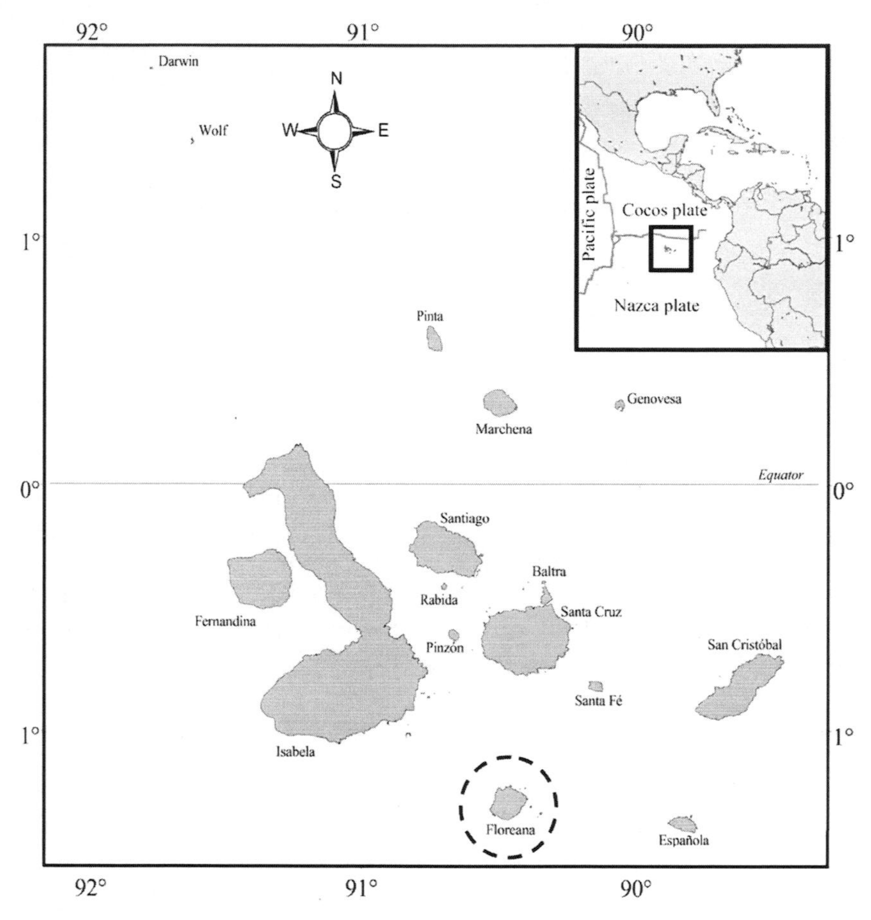

FIGURE 6.1 Location of the Galápagos archipelago. Insert shows the position of the islands in relation to the Americas and the lines show the fractures among tectonic plates of the eastern Pacific. Dotted circle indicates the position of Floreana Island where the food-web model described in this chapter was developed for the rocky reefs, which are typical of those found throughout the central shelf region of the archipelago.

formed 2.8–5.6 mya, and Fernandina as recently as 60–300 thousand years ago (Geist 1996, Geist et al. 1998). The west and southern sides of the Galápagos platform slopes to depths of >3 km about 50 km from the coast, which allows a bathymetric and geomorphologic deflection of undercurrent waters to the surface, creating persistent deep-water equatorial upwellings. The bathymetric gradient is more gradual to the northeast where the Cocos and Carnegie Ridges merge (Chadwick 2003).

Floreana Island is the sixth largest island and it lies along the south-central margin of the Galápagos platform (fig. 6.1). According to Bow and Geist (1992), Floreana shield development and lava flows have been dated to 1.1–0.77 mya with evidence of an eroded land bridge to a satellite island Champion. In this region there is considerable mixing of species that have colonized the island from various origins, including Chile-Peru, Mainland Ecuador, and central Pacific (James 1991). The central-south zone is the largest biogeographic region of the archipelago, and it contains many taxa and biotic components from those centers of origin (Edgar et al. 2004).

Climatic Setting

Despite the equatorial position of the Galápagos Archipelago, both tropical and temperate conditions exist. Two major seasons, the cold-dry period from June to November and the hot-wet period from December to April, are reflected by most environmental variables (fig. 6.2). Sea and air temperatures are closely correlated, as are rain fall and solar exposure at or near sea level. The air pressure differential between the Indonesian low and Pacific high-pressure systems drives wind predominantly to the west or northwest. To the north lies the low-pressure Intertropical Convergence Zone (ITCZ), which forms an extensive cloud belt, giving rise to tropical storms and rainfall that cross the heated islands as the southeast trade winds weaken between December and April. This change results in a net decrease of the atmospheric pressure, which in turn increases the sea surface height to its annual maximum around the months of April and May (fig. 6.2). During the cold-dry season (June–November), an inversion layer forms over the cooling sea surface and precipitates as a fine and persistent mist known locally as *garúa*. Ocean productivity also varies seasonally. The *garúa* season is characterized by cold, generally "blue-green" water with average visibility ranging from 6–9 m at the surface, while hot-wet months (December–May) bring lower productivity and a visibility range of 10–12 m, during which periods of exceptional visibility of greater than 20 m sometimes occur.

Oceanic Currents

The geographical setting provides the islands with a unique oceanographic environment: a tropical archipelago situated between major ocean currents and exposed to persistent upwelling conditions (Houvenaghel 1984, Chavez and Brusca 1991). The diversity of Galápagos marine habitats and communities is a reflection not only of the geology and varied oceanography at regional to ocean basin scales, but also of within- and between-year variability. The Galápagos are located at the confluence of three major oceanic currents that exhibit distinct seasonality (fig. 6.3). The Southern Equatorial Current is a blend of tropical and subtropical waters that generates net surface transport to the west, and these changes in intensity throughout the year. The Peru coastal current, also known as the Humboldt Current

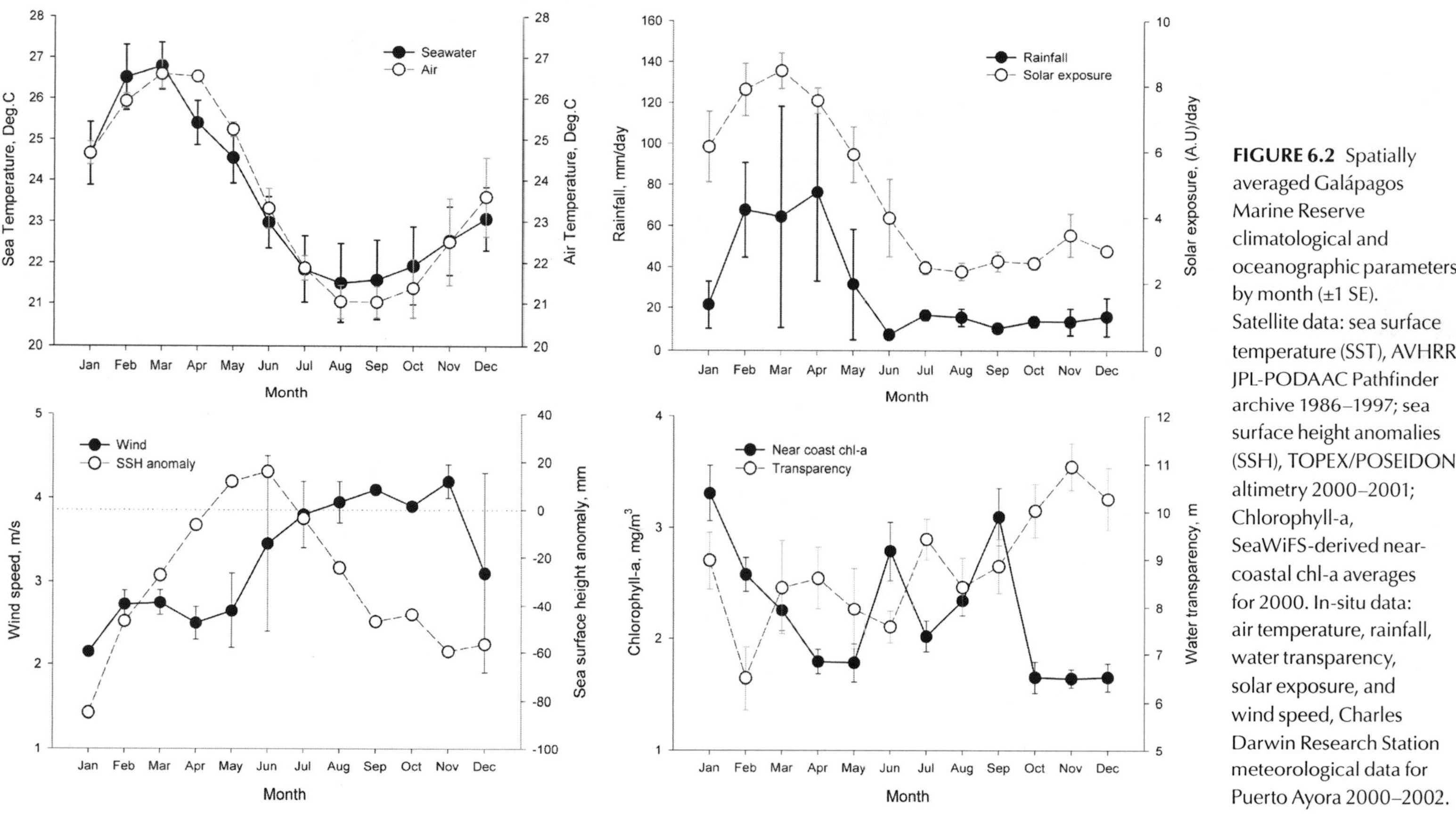

FIGURE 6.2 Spatially averaged Galápagos Marine Reserve climatological and oceanographic parameters by month (±1 SE). Satellite data: sea surface temperature (SST), AVHRR JPL-PODAAC Pathfinder archive 1986–1997; sea surface height anomalies (SSH), TOPEX/POSEIDON altimetry 2000–2001; Chlorophyll-a, SeaWiFS-derived near-coastal chl-a averages for 2000. In-situ data: air temperature, rainfall, water transparency, solar exposure, and wind speed, Charles Darwin Research Station meteorological data for Puerto Ayora 2000–2002.

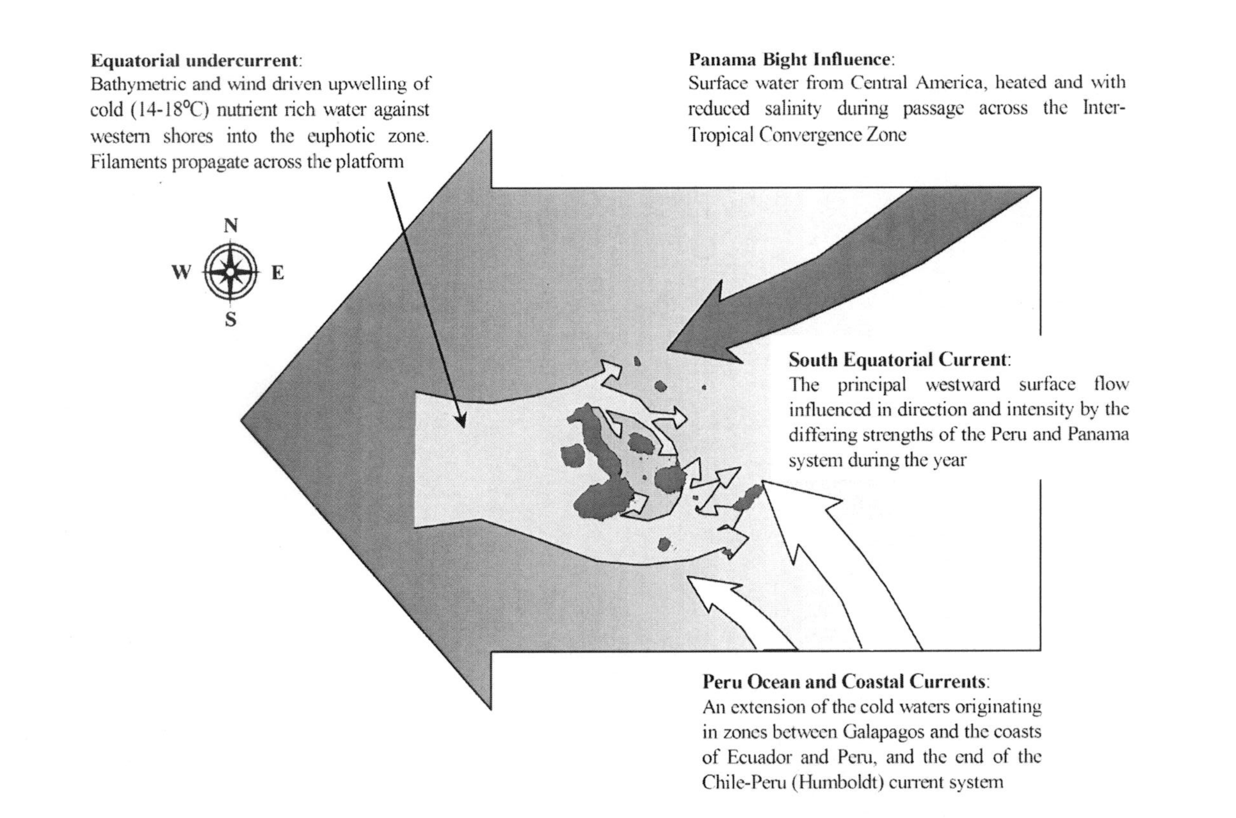

FIGURE 6.3 Schematic depiction of the major surface currents and water movement around the Galápagos Islands. White arrows from the south indicate the main cold-water flows (ocean and coastal Chile–Peru or Humboldt currents), and the dark grey arrow from the north show the warm waters (from north equatorial countercurrent). Light gray arrows from the west depict localized nutrient-rich cold upwelling cells of high productivity.

(18°C–24°C), together with the Peru oceanic current from the southeast, are the dominant currents during the cool *garúa* season of June to November. This flow from the south brings the majority of cold-adapted temperate species. The subtropical waters are warm, yet variable, and tend to be more saline (around 35% towards the equator) as a result of evaporation during passage in the south Pacific subtropical gyre. The westward advection of cooler surface water lowers temperatures locally (Wyrtki 1966). A combination of the extension of wind-driven equatorial upwelling at low latitudes that forms a cold tongue extending from the continent and entrainment of colder deep upwelled water advecting from the Peru coastal margin.

The Panama Bight Influence arrives from the northwest, and is an extension of the North Equatorial Counter Current that brings warmer and less saline tropical water, particularly during the wet season from December to June. During El Niño Southern Oscillation (ENSO) events, this current is believed to transport Panamic and even Californian species to the Galápagos, particularly to the more tropical assemblages of the northern and central islands. The salinity of this current spans 30%–34% depending on the amount of rainfall. Temperature and salinity differences between the Humboldt and Panama Flows produce an oceanic front at the confluence of the two water masses creating temperature gradients of as much as 5°C over 50 km. During the austral summer (December–May) the Panamanian Bight Influence descends from the north, homogenizing temperatures across the archipelago. Both of these surface currents are relatively impoverished in nutrients after extended circulation in open ocean gyres.

Soluble iron is upwelled from bathymetric deflection of the eastward flowing submarine Equatorial Undercurrent (also known as the Cromwell Current) against the Galápagos platform. Iron appears to be a limiting micronutrient involved in the uptake or chelation of nitrates (Coale et al. 1998) in what is otherwise recognized as a high-nitrate and low-chlorophyll area. The Equatorial Undercurrent is normally positioned some 100 m below the surface, well below the euphotic zone (Wyrtki 1985), where it entrains recycled nutrients from the upper ocean as it meanders from the central equatorial Pacific and propagates across the Galápagos platform. This produces zones of persistent nutrient-enriched upwelling at the western shores of most Galápagos Islands before reforming some 100 km to the east of the archipelago. In addition to supporting immense phytoplankton blooms spanning hundreds of kilometers, it generates areas of consistently cooler water (≈12°C –18°C) allowing species to exist in the Galápagos that would otherwise not be found at the equator, including penguins and fur seals. Extensive temperature and primary production anomalies spanning areas of more than 200 km^2 can evolve and disappear in less than two weeks.

Marine Habitats

Complex current dynamics are largely responsible for the sporadic character of colonization by marine biota from disjunct biogeographic regions, which led to the evolution of unique species. Galápagos marine habitats above the central-south shelf can be divided into two realms: (1) the shallow coastal and benthic inshore areas, and (2) the deeper open pelagic areas (Bustamante et al. 2002a). The food web described in this chapter is composed mostly of biota that inhabits the coastal and shallow benthic habitats with some overlapping pelagic biota of the south-central islands region. The modeled food web also includes several "pelagic" functional groups, including sharks, pelagic fishes, cetaceans, seabirds, sea turtles, and plankton, which visit or inhabit the coastal lava reefs.

Most of the coastal shores of the Galápagos consist of consolidated and sloping lava fields. More than 90% of the shallowest benthic habitats are lava reefs with interspersed sand pockets composed of both biogenic material (formed by corals and echinoid tests) and pulverized lava that forms brown, red, and black sand. Lava reefs are present around all islands and are interspersed with other habitats such as vertical walls, sandy beaches, and mangroves. Coral reefs are the rarest of all shallow habitats in the Galápagos, being restricted presently to a few patches of several hundred square meters at a few islands such as Darwin, Wolf, Española, and Genovesa. Floreana Island had large patches of corals reefs that were well studied in the past (Glynn et al. 1979, Wellington 1984) but were decimated by the 1982–1983 El Niño (Robinson and Del Pino 1985, Glynn 1988). However, remnant patches and coral colonies are found interspersed at Floreana Island and its nearby islets of Champion and Enderby. These remnant habitats are declining rapidly at most remaining locations due to successive ENSO stresses and grazing by dense populations of sea urchins and fish (Glynn et al. 1979; Glynn 1990, 1994).

The Biota

Biogeography and Biodiversity

Interactions of cold and warm currents in this highly complex and dynamic Galápagos environment have produced several discrete biogeographic zones separated by short distances (Abbott 1966, Harris 1969, Glynn and Wellington 1983, James 1991). Between three and five major biogeographic units have been proposed for the archipelago. Harris (1969) proposed five units based on his study of seabird nesting and distribution and sea temperature. His 35-year-old biogeographic divisions still appear valid for the marine ecosystems, but the number of units and their boundaries require refinement (Jennings et al. 1994, Bustamante et al. 2002b). Edgar and colleagues (2004) reviewed this model and suggested three large biogeographic bioregions: the western temperate-cold zone, the far northern tropical-warm zone, and south-central/eastern mixed temperate-subtropical, based on benthic mobile macro-invertebrates and reef fishes. The western and south-central/eastern bioregions can also be further sub-divided into two smaller and nested regions (Edgar et al. 2004). These large biogeographic bioregions are identified by their distinctive and unique biota, which have colonized from four sources: the Indo-Pacific, the mainland and Peruvian, the Panamic-Caribbean, and the local endemic source areas (McCosker and Rosenblatt 1984, Kay 1991, James 1991, Edgar et al. 2004).

The isolation of the Galápagos Islands has led to a diverse biota with a high proportion of endemic marine species. To date, 2,909 species of marine mammals, macroalgae, marine birds, fish and invertebrates have been recorded in the Galápagos with an overall endemism of 19% (Bustamante et al. 2000b). Colonization and speciation has varied greatly among taxonomic groups. Mollusks, fish, and algae, for instance, are highly diverse, while other groups such as barnacles, gorgonians, and porcelanid crabs are species-poor. The average endemism is 26% for the 16 taxonomic groups (fig. 6.4). Relative to the more speciose taxa, the less speciose groups tend to exhibit higher endemism and higher variability of endemism. Several highly charismatic Galápagos endemics, such as the marine iguana, the flightless cormorant, and the Galápagos penguin depend on the high level of marine production. Although Galápagos marine communities exhibit high endemism, it is considerably lower than in terrestrial Galápagos communities where, for example, all reptile and mammal species are endemic and nearly a third of all vascular plants are endemic (Tye et al. 2002).

The Food Web

An explicit mass-balance model for the shallow rocky reef ecosystem was constructed to represent the state of the food web around the year 2000 in waters shallower than 20-m depth in the northern portions of Floreana Island. This particular lava reef ecosystem was chosen, in part, because it offered the best ecological data at the time and represents the Galápagos south-central shelf region in that it harbors a mix of species representative of temperate and subtropical south-central and eastern bioregions (Edgar et al. 2004). The model was constructed collaboratively by Okey and colleagues (2004) using the Ecopath and Ecosim modelling approach (Christensen and Walters 2004) to integrate information from an extensive baseline ecological monitoring program for the Galápagos Marine Reserve, related field programs undertaken by the Charles Darwin Research Station and partners, the scientific literature, local knowledge of the system, and other sources.

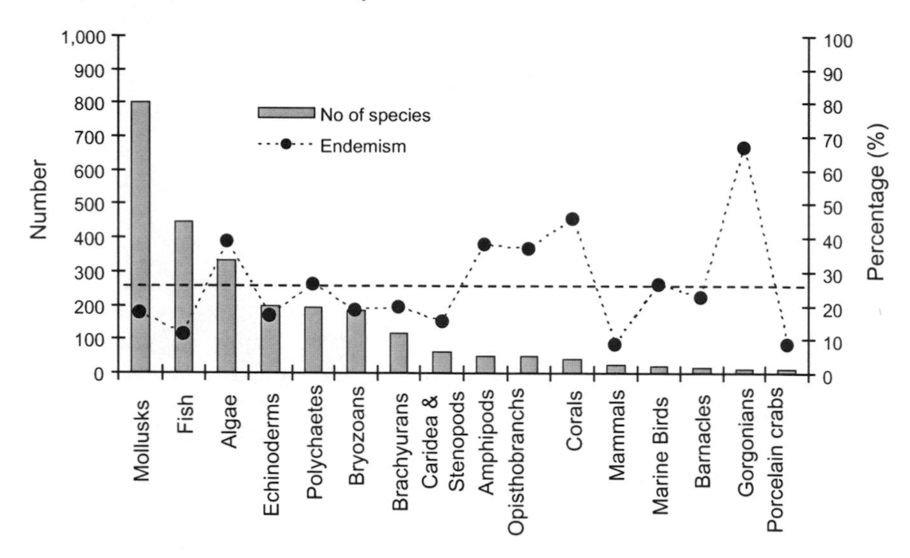

FIGURE 6.4 Species richness and marine endemism of major groups recorded in the Galápagos Islands. Horizontal dotted line indicates the 26% average level of marine endemism among these taxonomic groups.

The basic building blocks of Ecopath models are estimates of biomass, production, consumption rates, and diet for each defined functional group within a defined model area and time. The resulting flow characteristics are used as starting point for analyzing temporal and spatial dynamics scenarios to assess changes in exploitation, resource policies, the environment, or ecological trophic interactions. In total, 43 functional groups were defined to represent the local biodiversity of this food web, including both benthic and pelagic components (tab. 6.1). These functional groups range from aggregations of a wide range of trophically similar species, such as "small crustaceans," sessile "filter and suspension feeders," and pelagic predatory fishes with 11 taxa, to single-species groups such as the pencil sea urchin (tab. 6.1). Aggregations were based on similarities of diets and functions, but also on strength of interactions and utilitarian interests, notably the sea cucumber *Stichopus fuscus*.

In addition to summarizing the rocky-reef food web diagrammatically and with basic descriptive statistics, the 43-groups composing the model food web into five descriptive trophic-level (TL) categories spanning trophic levels 1.0 to 4.4: primary producers and detritus (TL = 1.0), primary consumers (TL = 2.0), mixed primary consumers (TLs 2.1–2.4), secondary consumers (TLs 2.5–3.4), and top predators (TLs 3.5–4.4). The largest total biomass per TL was contributed by primary producers followed by mixed consumers, while lesser total biomass levels were recorded for both primary and secondary consumers and, as expected, the lowest biomass was recorded for top predators (fig. 6.5).

Table 6.1. Trophic levels, functional groups, indices, and taxa included in the Galápagos food-web model.

	Group name	Trophic level	Biomass (t km^{-2})	ISI	Keystone index	Taxa
Top predators	Sharks	4.4	0.75	8.6	286.7	*Carcharhinus galapagensis, Carcharhinus limbatus, Sphyrna lewini, Trianedon obesus*
	Toothed cetaceans	4.4	0.02	1.6	1600	*Tursiops truncatus, Orcinus orca, Pseudorca globicephala*
	Bacalao grouper	4.2	7.14	1	3.6	*Mycteroperca olfax*
	Birds	4.1	0.01	0.2	575	*Sula nebouxi, Sula dactylatra, Pelecanus occidentalis, Spheniscus mendiculus*
	Sea lions	4	5.68	9.4	42.8	*Zalophus californianus wollebaki*
	Pelagic predatory fishes	3.9	30	22.4	19.6	*Acanthocybium solandri, Thunnus albacares, Seriola rivoliana, Coriphaena sp., Katsuwonus* sp., *Scomberomorus sierra, Sphyraena idiastes, Euthynnus linneatus, Sarda orientalis, Elagatis bipinnulata, Trachinotus stilbe*
	Non-commercial reef predators fishes	3.8	14.86	11.5	20.2	*Cirrhitus rivulatus, Hemilutjanus macrophthalmus, Fistularia commersoni, Gymnothorax dovii, Gymnothorax castaneus*
	Octopus	3.5	0.79	0.9	30	*Octopus* sp.
	Total Biomass = 59.25					
	Pelagic planktivores fishes	3.4	5.5	1.6	7.6	*Opisthonema* sp., *Sardinops sagax, Xenocys jessiae, Hyporhamphus unifasciatus*
	Other commercial reef predators fishes	3.3	9.3	1.7	4.9	*Paralabrax albomaculatus, Epinephelus panamensis, Dermatolepis dermatolepis, Epinephelus labriformis, Cratinus agassizi, Lutjanus aratus, Lutjanus argentiventris, Lutjanus jordani, Hoplopagrus guentheri, Lutjanus novemfasciatus, Lutjanus viridis*
	Large benthic invertebrate eaters fishes	3.3	32.71	18.7	15	*Bodianus diplotaenia, Bodianus eclancheri, Semicossyphys darwin, Halichoeres nicholsi, Balistes polylepis, Sufflamen verres, Pseudobalistes naufragium, Heterodontus quoyi*

(Continued)

Table 6.1. (Continued)

Group name	Trophic level	Biomass (t km^{-2})	ISI	Keystone index	Taxa
Planktivorous reef fish	3.3	281.13	7	0.7	*Paranthias colonus, Apogon atradorsatus, Abudefduf troscheli, Myripristis berndti, Myripristis leiognathos, Chromis alta, Chromis atrilobata*
Conch gastropod	3	3.61	3.7	26.4	*Hexaplex princeps, Fasciolaria princeps*
Small benthic invertebrate eaters fishes	3	100.99	13.4	3.5	*Halichoeres dispilus, Thalassoma lucasanum, Holocanthus passer, Haemulon scudderi, Haemulon sexfasciatum, Orthopristis forbesi, Anisotremus interruptus, Anisotremus scapularis*
Carnivorous zooplankton	2.8	3.58	8.8	n/a	Undetermined
Spiny lobsters	2.8	3	2.6	23.6	*Panulirus gracilis, Panulirus penicillatum*
Slipper lobster	2.7	4	0.7	4.7	*Scyllarides astori*
Omnivorous reef fishes	2.7	41.52	17.7	11.2	*Stegastes arcifrons, Stegastes leucorus beebei, Chaetodon humeralis, Johnrandallia nigrirostris, Zanclus cornutus, Girella freminvillei, Kyphosus elegans, Kyphosus analogus*
Shrimps and small crabs	2.6	55.13	18	8.6	Shrimps, crabs, stomatopods
Asteroids	2.5	10.49	0.4	1	*Nidorellia armata, Pentaceraster cumingi, Phataria* sp., *Lynkia* sp.
Total Biomass = 550.96					
Other herbivorous fishes	2.4	200.6	13.4	1.7	*Prionurus laticlavius, Microspathodon dorsalis, Microspathodon bairdi*
Pencil urchin	2.2	104.43	8.4	2.1	*Eucidaris thuarsii*
Anemones	2.2	79.24	3.1	1	*Actynaria* sp., *Bunodactys* sp., *Anthopleura* sp., *Aptasia* sp.
Worms and ophiuroids	2.2	84.67	10.3	3.2	*Ophiura* spp., *Annelida, Sipunculida*

(Continued)

Mixed primary consumers	Haermatypic corals	2.2	91.16	2.6	0.7	*Pocillopora* spp., *Psammocora* sp., *Porites lobata*, *Pavona clavus*
	Chitons	2.2	2.85	0.1	0.9	*Chiton* sp. *Tonicia* sp., *Acanthochiton* sp.
	Detritivorous fish	2.1	39.95	0.6	0.4	*Mugil* spp. *Chanos* spp.
	Small gastropods	2.1	188.05	6.4	0.9	*Cerithium adustum*, Planaxidae, Ranelidae, *buras*, *corrugata*
	Sea turtles	2.1	3.02	0.2	1.7	*Chelonia mydas*
	Pepino sea cucumber	2.1	3.9	n/a	n/a	*Stichopus fuscus*
	Total Biomass = 797.87					
Pure primary consumers	Other urchins	2	4.65	0.1	0.6	*Echinometra vanbruntii*, *Centrastephanus coronatus*, *Diadema antillarum*
	Parrotfishes	2	21.5	1.7	2.1	*Scarus ghobban*, *Scarus perrico*, *Scarus rubroviolaceus*, *Scarus compressus*
	Marine iguana	2	0.8	0.1	3.3	*Amblyrrhinchus cristatus*
	Other sea cucumbers	2	3.55	0.1	0.7	*Stichopus horrens*, *Holothuria arenicola*, *Holothuria difficilis*, *Holothuris impaties*, *Holothuria leucospilota*
	White urchin	2	48.74	3.9	2.1	*Tripneustes depressus*
	Green urchin	2	8.72	0.5	1.5	*Lytechinus semituberculatus*
	Small crustaceans	2	91.41	0.5	0.1	Amphipods, isopods, tanaids, mysids, other pericarids
	Filter + suspension feeders	2	367.39	9	0.6	Cirripedia, *Pacifigorgia* sp., *Tubastrea coccinea*, bryozoans, Ascidia, Porifera
	Herbivorous zooplankton	2	3.19	10.9	n/a	Undetermined
	Total Biomass = 549.95					
Primary producers	Phytoplankton	1	12	3.7	n/a	Undetermined
	Microphytobenthos	1	393.59	16.1	1.1	Undetermined
	Benthic algae	1	256.8	16.5	1.7	Undetermined
	Detritus	1	500	n/a	n/a	n/a
	Total Biomass = 1162.39					

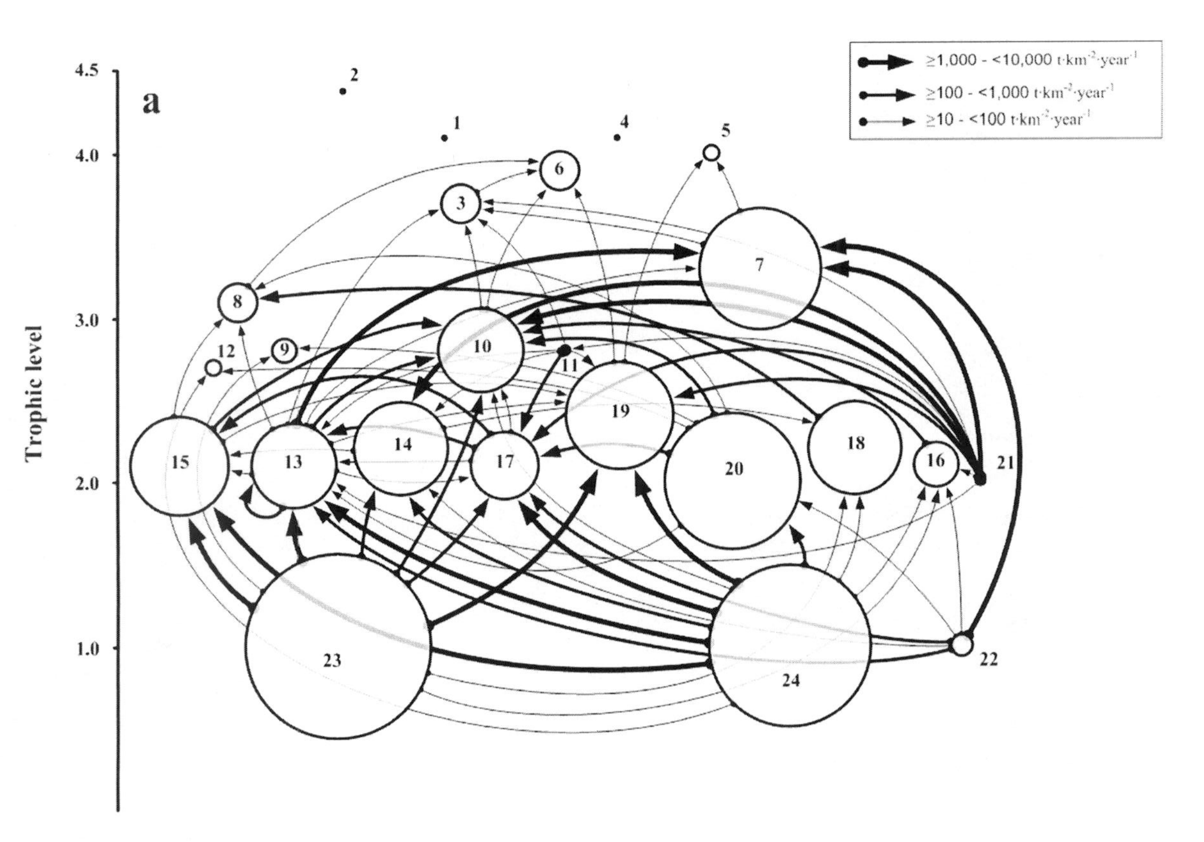

FIGURE 6.5 (Continued).

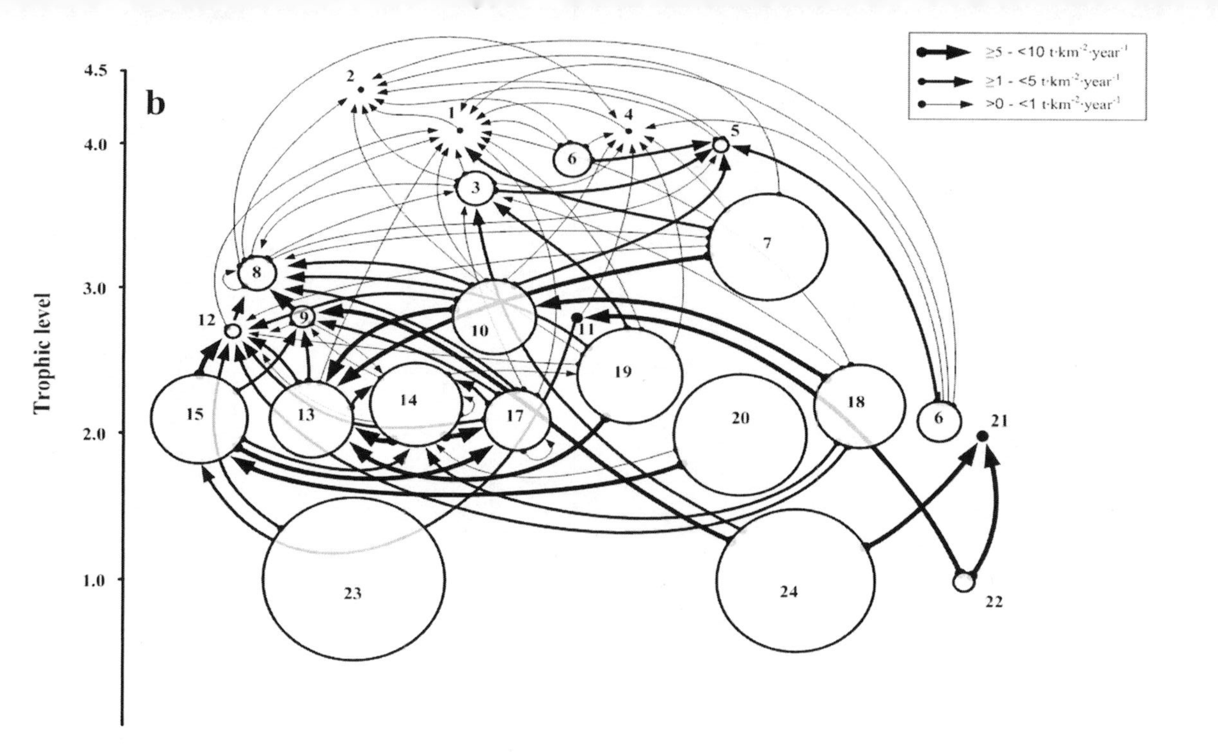

FIGURE 6.5 Simplified food-web diagram for the Floreana Rocky Reef, Galápagos. Circles indicate each functional group, while numbers indicate their identity (see below) and the size indicates a proportional scaling of their relative biomass among each group. Arrows indicates the direction of the biomass flow from prey to predator. Panel (a) depicts 74 connecting lines for those biomass flows that range from 10–10,000 $t \cdot km^{-2} \cdot year^{-1}$ and (b) depicts 83 biomass flows that range between 0.1– 10 $t\ km^{-2}\ year^{-1}$. Numbers are: 1. pelagic sharks, 2. dolphins, 3. large reef predatory fishes, 4. seabirds, 5. sea lion, 6. pelagic predatory fishes, 7. pelagic planktivorous fishes, 8. large benthic predatory fishes, 9. macro-invertebrate carnivores, 10. small omnivorous reef fishes, 11. carnivorous zooplankton, 12. macro-crustacean carnivores, 13. micro-crustacean carnivores, 14. hard corals and zoanthids, 15. micro-gastropods grazers, 16. detritivores fishes, 17. benthic invertebrate detritivores, 18. macro-invertebrate grazers, 19. vertebrate grazers, 20. filter and suspension feeders, 21. herbivorous zooplankton, 22. phytoplankton, 23. benthic macroalgae and microphytobenthos, 24. detritus. Circles in white and grey indicate benthic and pelagic functional groups, respectively.

The biomass of primary producers was dominated by a large component of detrital material that was either imported from the pelagic or generated by local decay and waste of benthic species. Microphytobenthic species, mostly benthic diatoms, and abundant macroalgal stands are also major contributors to benthic primary production. Thus, phytoplankton was not one of the major primary producers within this system, though it is continually imported. Primary consumers were largely dominated by a large biomass of suspension and filter feeder species such as barnacles, gorgonians, sponges, ascidians, and bryozoans (tab. 6.1). In addition, they included a large component of small grazing gastropods such as *Cerithium adustum*, Planaxidae, Ranelidae, *Bursa corrugata*, and sea urchins, in particular the white and green sea urchins, *Tripneustes depressus* and *Lythechinus semituberculatus*. The high biomass of the mixed consumer functional group was composed mainly of herbivorous species, partial producers such as hermatypic corals, but also omnivorous species, including many species of fish, sea turtles, small grazing-scavenging gastropods, and in particular the omnivorous pencil sea urchin, *Eucidaris thouarsii*.

The total biomass of the secondary consumers was largely accounted for by two functional groups, the planktivorous fishes, such as the Pacific creole fish *Paranthias colonus*, and abundant small and large predatory fishes that prey on benthic invertebrates, such as the wrasses *Halichoeres dispilus, Thalassoma lucasanum, Holocanthus passer,* the triggerfishes *Sufflamen verres, Pseudobalistes naufragium, Balistes polylepis*, the schooling grunts *Haemulon scudderi, Haemulon sexfasciatum,* and the large wrasses such as *Bodianus diplotaenia, Bodianus eclancheri, Semicossyphus darwini,* and *Halichoeres nicholsi.*

The biomass of top predators was accounted for largely by a mix of different pelagic and benthic vertebrates, including fishes, mammals, seabirds, and sharks, and the Galápagos octopus *Octopus oculifer.* The largest components were a broad mix of predatory pelagic species of fishes such as the wahoo, *Acanthocybium solandri,* the tunas *Thunnus albacares, Euthynnus linneatus, Sarda orientalis,* and *Katsuwonus pelamis,* the Pacific amberjack *Seriola rivoliana,* the mahi-mahi *Coriphaena hippurus,* the Sierra *Scomberomorus sierra,* and the Pacific barracuda *Sphyraena idiastes*. Other major components are the Galápagos grouper *Mycteroperca olfax*, a key target of the local fishing, followed by fluctuating populations of the Galápagos sea lion *Zalophus californianus wollebaki.*

Simplifying the Food Web

The modelled food web of 43 functional groups described above had a total of 299 trophic connections that are difficult to represent in a classical two dimensional food-web diagram. Consequently, we constructed a simplified web by further pooling ecologically similar trophic groups, for example spiny lobsters and slipper lobster, all sea urchins, and all sea cucumbers. The resulting diagrams included 24 functional groups with a total of 157 connections. Due to the still complex number of connections for a single food web, the diagrams were separated according to the magnitude of biomass flows (fig. 6.5). The arrows in the figures depict biomass flows from prey to predator, circles indicate the proportional biomass of each functional group; and the shaded circles indicate pelagic components.

The largest biomass flows range between 10 and 10,000 t·km^{-2}·year^{-1} (fig. 6.5a) while the smaller flows range between 0 and 10 t·km^{-2} year^{-1} (fig. 6.5b). For each food web diagram, the biomass flows were further divided into three size classes depicted with different arrow thickness. These show that the largest flows were between primary producers and their consumers and the smallest among consumers. In contrast to the magnitude of flow, the number of trophic connections reaches its peak with the smallest biomass flows (10–100 t·km^{-2}·year^{-1}), which account for 50% of the total 74 connecting flows. Only 16% of all biomass flows reach the top predators. These flows are primarily the consumption of mixed and secondary consumer species, such as small pelagic planktivorous, and omnivorous reef fishes eaten by large reef fishes, sea lions, and pelagic predatory fishes (fig. 6.5a).

The food web for low-magnitude flows shows that of its 14 larger biomass flows (5–10 t·km^{-2}·year^{-1}), only four occur from detritus and phytoplankton to consumers (fig. 6.5b), and none from this range reached the top predators. As would be expected, intermediate biomass flows in this lower flows diagram (1–5 t·km^{-2}·year^{-1}) largely originated from mixed consumers and secondary consumers, while the smallest biomass flow (0.1–1 t·km^{-2}·year^{-1}) originated mostly from consumers to the top predatory groups. This analysis shows that the top predators accounted for more than 66% of the total number of the smallest biomass flows (0.1–1 t·km^{-2}·year^{-1}) and more than 60% of the total number of trophic connections, reflecting the broad diet of Galápagos top predators. In spite of this broad diet and relatively small biomass flow, upper trophic level predators appear to shape Galápagos marine ecosystems strongly (Vinueza et al. 2006) and their role would be even stronger if they were not so depleted.

Food Web Properties and Dynamics

The Galápagos rocky-reef food web is a net importer of food, as indicated by a highly negative net system production estimate of –14,300 tons wet weight·km^{-2}·year^{-1} and negative export estimate of –5400 t·km^{-2}·year^{-1}. This property emerges despite the high estimates for primary production on this shallow rocky reef food web of 13,300 t·km^{-2}·year^{-1}. Net heterotrophy is probably common for the food webs of most oceanic reef ecosystems because of high oceanic inputs. That is to say, the high abundances of biota at oceanic island ecosystems and seamounts cannot be explained without considering trophic imports from the large primary production in the surrounding oceanic ecosystem, which becomes concentrated around these features. The estimated overall respiration was twice the total primary production. Still, the ecosystem's annual primary production was five times the average standing biomass of 2600 t·km^{-2}, indicating a high turnover. The sums of all production, consumption, respiration, and flows to detritus were 17,300, 51,600, 27,600 and 21,000 t·km^{-2}·year^{-1}, respectively, and the proportion of trophic flows originating from detritus was 0.62.

We defined the trophic level categories in order to characterize the food web with regard to the trophic distributions of some informative food web properties (tab. 6.2). The first of the three distributions is the relative distribution of biomass among the five feeding categories (fig. 6.6). Aquatic ecosystems are often thought to have a low biomass, but a high turnover, of primary producers. This is true of fast-growing plankton that supports much larger standing biomasses of higher trophic level organisms. Both the biomass and the turnover of primary producers were high as a consequence of both benthic frondose algae and microphytobenthos being present. As with any ecosystem model, there is uncertainty in our estimates, but the biomass of benthic primary producers was certainly high compared with other estimates and models of marine ecosystems.

Two indices of food-web dynamics from Okey (2004)—the Interaction Strength Index (ISI) and the Keystone Index—were

Table 6.2. Flows from primary production and detritus in the Ecopath model of Floreana rocky reef, Galápagos.

	From Primary Producers		From Detritus	
TL	Export	Throughput	Export	Throughput
VI	0	1	0	1
V	0	20	0	16
IV	0	621	0	561
III	–927	5,533	–927	2,855
II	–8,843	24,448	–3,648	17,544
I	–16,00	13,223	10,534	21,124
Sum	–11,370	43,846	5,959	42,102

Flows are expressed in tons wet weight·km^{-2}·year^{-1}. Some flows reach trophic level VI because some species are supported by energy that has traversed five links from primary producers.

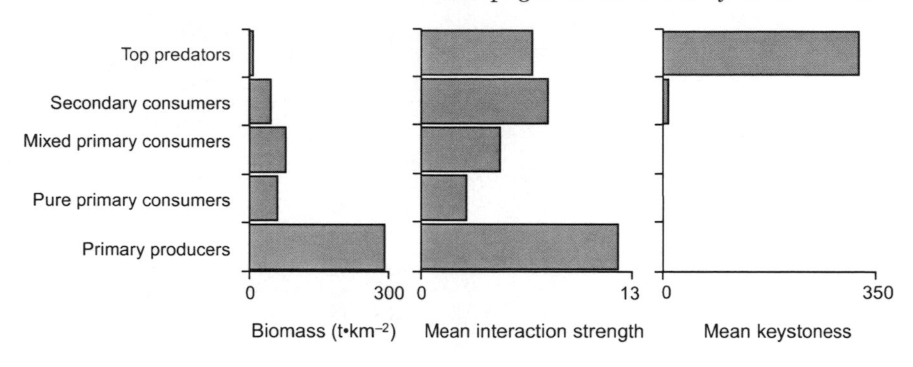

FIGURE 6.6 Distributions of three trophic characteristics among the five delineated feeding-level categories in the Ecopath model of Floreana rocky reef, Galápagos.

calculated for the Galápagos rocky reef ecosystem by conducting virtual simulations of functional group removals from the 43-group food web. The Interaction Strength Index is the sum of the predicted change in all the "affected" groups 25 years after simulated removal of the affecting group in question. The Keystone Index is a group's ISI divided by the relative biomass of the removed group. Means for interaction strengths and keystone indices were estimated for each broad trophic category (fig. 6.6). Unlike pelagic systems, a large bulk of the standing biomass of this reef system comprised primary producers (macroalgae) and detritus. But perhaps like all marine ecosystems, interactions strengths were more equally distributed and mean "keystoneness" was concentrated at the highest trophic levels. Since keystone species affect the structure, function, and diversity of biological communities in a way that is disproportionately strong relative to a group's biomass (Paine 1969, Bond 1993, Power et al. 1996), their loss might radically change the structure and function of the whole ecosystem. The Galápagos is experiencing such losses.

The largest number of functional groups and the greatest biomass were found among primary and mixed consumers, at around trophic level 2 (fig. 6.6). There are two possible explanations for this unusually high abundance of intermediate consumers. The first is that it results from high production and biomass of benthic primary producers and high imports of oceanic plankton and detritus. This explanation is likely to be too simplistic because top-level consumer fauna would be expected to immigrate and diversify in response to this food source, which does not appear to have occurred. The second explanation is that predators have been depleted and this has released prey leading to high biomass, such as observed for filter and suspension feeders, sea urchins, and sea cucumbers. This explanation seems more likely, especially given the history of sequential depletions of high level

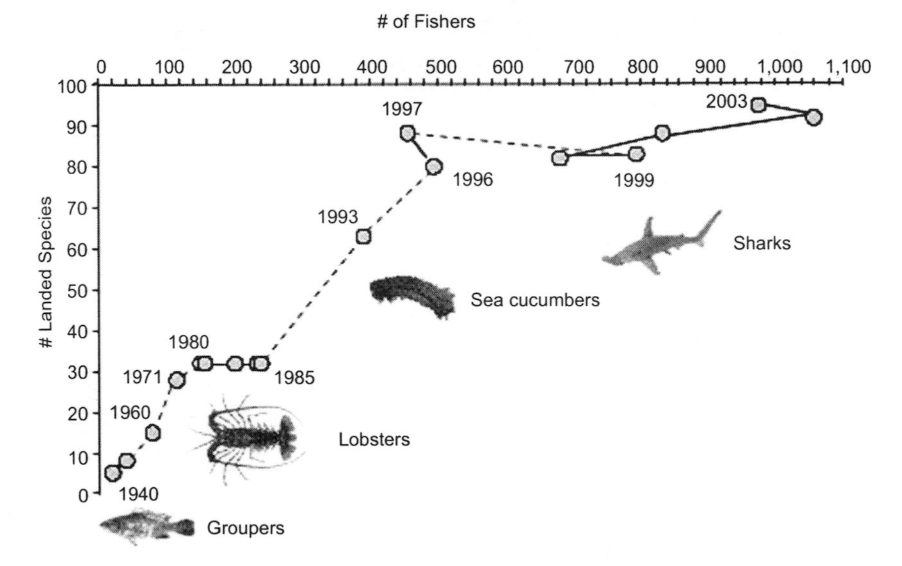

FIGURE 6.7 Historical changes in the exploitation of the Galápagos food web. Circles indicate the numbers of landed marine species in relation to the numbers of local fishers in the Galápagos archipelago. Pictures indicate the main target species. Dotted lines indicate discontinuities in the time series. Data for the fishers between 1940 and 1960 were estimated from personal interviews with older fishers and residents.

consumers and top predators in Galápagos by artisanal fisheries (fig. 6.7) and the observed ecological changes related to fisheries, such as the sea urchin-coralline barrens. Consequently, the food web presented here has probably already been significantly affected by serial fisheries extractions.

Human Influences

Waves of Exploitation

Galápagos marine ecosystems have been influenced by successive waves of human exploitation, starting approximately 400 years ago when giant land tortoises were collected by passing ships (Beebe 1924). From the late 1700s to mid 1800s commercial whaling initiated another level of exploitation that ended due to the scarcity of whales and the replacement of whale oil lamps with kerosene lamps (Perry 1984 ; Jackson 1994; Whitehead et al. 1997; Merlen 1992, 1995). Whalers also discovered that the islands had fur seal populations, which they quickly reduced to near biological extinction (Jackson 1994, Roberts 1999). The short-lived "boom-bust" nature of whaling and sealing foreshadowed what was to come.

Exploitation has increased step-wise over the last 60 years. Successive waves have been characterized by increases in the numbers of fishers and their fishing vessels, as well as increases in catches and the diversity of target species. These changes, combined with advances in fishing technologies, have progressively increased the accumulated fishing effort (fig. 6.7). Pulses of exploitation have occurred at irregular intervals due to the arrival of new and more profitable fishing practices, especially diving and the exploitation of new target species. New markets emerged for these species because of declines in the production and profitability of previously targeted stocks, internationalization of seafood markets, and the increased buying power of Asian markets, all these factors being new to Galápagos.

Local artisanal and small-scale fishing began when Ecuadorians colonized the archipelago during the late 1800s, but there was probably not much fishing until the early 1900s (Bustamante et al. 2002c). Since then, at least three successive and cumulative waves of exploitation have occurred. In the early 1920s, a group of Norwegians initiated commercial fishing for groupers, but only in the 1940s to 1950s did commercial artisanal fishing start, beginning on a small scale involving fewer than 30 fishers (fig. 6.7) from a total island population of approximately 5000 people (Hoff 1985, Reck 1983). This first wave of fishing solely targeted reef-dwelling groupers *Mycteroperca olfax* and about three species of deep-water groupers of the genus *Epinephelus*, almost exclusively for national fresh and salt-dried markets (Reck 1983). From the 1960s to the late 1980s, financial incentives from international fish prospectors produced a rapid development of a second wave of fishing, including lobster fishing using a surface-supply diving technique, namely the hookah (Bustamante et al. 2002d). The third and most recent and substantial wave of exploitation emerged during the early 1990s with the introduction of sea cucumber fishing (Richmond and Martinez 1993, Bustamante et al. 2002c), which was built upon lobster diving technology, and dramatically changed the fishery and food web.

After a 4-year moratorium, for a 2-month season for sea cucumber fishing was reopened in 1999. As a direct result, the number of fishers and the fleet size respectively increased by 92% and 54% from 1998–2000. In parallel, the number of the other targeted and incidentally caught species also rose dramatically, although in low catch volumes. By 2002, there were nearly 1000 fishers with more than 400 active fishing vessels ranging from 3–18 m lengths that targeted up to 95 different species of fishes and invertebrates (Bustamante et al. 2002c, Murillo et al. 2003). A new wave of fisheries exploitation is poised to break across the Galápagos archipelago. Offshore and deep water long-line fishing has been proposed by local fishers and has found support from Ecuadorian fishing companies, local politicians, and investors. These interests have proposed the introduction of off-shore long-line fishing, mainly for tuna and billfishes on a semi-industrial scale, which they advocate as a more acceptable alternative to the continually increasing fishing pressure on traditional coastal targets, such as the depleted lobsters, groupers, mullets, and sea cucumbers. Long-line fishing is currently conducted illegally in the Galápagos Marine Reserve, but if legalized this fishery would add to the intense pressure on large predatory fishes that international fleets have fished throughout the broader region for decades (Punsly 1983, Uosaki and Bayliff 1999, Okamoto and Bayliff 2003). This will add another case to the series of well-known adverse effects of fishing on marine ecosystems throughout the world (Myers and Worm 2003), this time to the top pelagic fish predators of the Galápagos marine ecosystem.

Effect on the Food Web and Biodiversity

These successive waves of exploitation of the food web have resulted in increasing total catches yielding an annual average of 2000 ± 300 metric tons wet weight of various species between 1999 and 2003, mostly for export to international markets. Sea cucumbers account for more than 70% of the total catches (tab. 6.3). To date, no moratoriums of targeted fish species have been implemented during the >60 years of continuous exploitation (although a 4-year moratorium was imposed between 1994 and 1999 for sea cucumbers). These three phases of fishing growth have resulted in serial depletion of the marine biota, including key components of both high and low trophic levels (fig. 6.7). The total estimated fishery extraction from the system is 4.2 $t \cdot km^{-2} \cdot year^{-1}$ and the mean trophic level of the catch is 2.3.

Table 6.3. Percentage of total annual catch comprising the 10 functional groups targeted in Floreana reef fisheries from 1997 to 2000.

Functional Group	Percentage of Total Catch	TL
Pepino (*S. fuscus* sea cucumber)	71.3	2.1
Detritivorous fishes (mullets)	15.2	2.1
Pelagic predators	5.4	3.9
Spiny lobster	4.3	2.8
Noncommercial predatory reef fishes	1.6	3.8
Other commercial predatory reef fishes	0.9	3.3
Bacalao grouper	0.8	4.2
Slipper lobster	0.3	2.7
Pelagic planktivore fishes	0.1	3.4
Large benthic invertebrate-eating fishes	0.1	3.3

TL, Trophic level.

This mean trophic level is very low compared with other systems and is due to large catches of sea cucumbers (*Stichopus fuscus*) and mullet species (*Mugil cephalus* and *M. rammelsbergii*), which graze on detritus, microbes, and phytoplankton. It is likely that these serial depletions have triggered considerable structural and functional changes on nearshore reefs through trophic cascades (Ruttenberg 2001, Okey 2004, Okey et al. 2004).

The impending large-scale, offshore long-line fishing proposed for the archipelago is likely to greatly increase the bycatch, which is currently low in the existing hand-line and dive fisheries. There is already substantial evidence that the use of long-lines in the greater region exerts massive and indiscriminate incidental killing and bycatch of sharks, sea turtles, sea lions, sea birds, dolphins, and many other predatory fish species (Belda and Sánchez 2001, Francis et al. 2001, Majluf 2002), most of which belong to higher trophic levels and have high interaction strengths and keystone effects.

Sharks are presently the biggest economic lure for local fishers as their fins fetch large sums of money from eastern Asia. This lucrative fishery is illegal in the Galápagos but has nevertheless expanded considerably in recent years due to its profitability (Zarate 2002). The depletion or removal of top predators such as sharks from Galápagos ecosystems will likely have strong effects on the overall structure and function of communities (Okey et al. 2004) since sharks are keystones of the Galápagos food web. The implications of this for the lucrative tourist trade are considerable, as the charismatic marine megafauna constitute an important part of what draws tourists to the region.

Conclusions

Our analysis of the current food web model of the Floreana rocky reef indicates that the overall web is strongly heterotrophic despite the relatively high standing-benthic biomass and production rates. Second, the state of the food web and the historical patterns of extractions indicate that the system has been modified strongly by fishing. This fishery is a good example of serial depletion, where initial exploitation of high trophic levels has replaced by current fishery targets at much lower trophic levels (Pauly et al. 1998). The most recent focus on targeting sharks and migratory pelagic species might reverse the "fishing down the food web" trend once again as effort is expanded to a larger area. Recent extraction rates of the main fishery targets—lobsters and sea cucumbers—are unsustainable (Okey et al. 2004, Shepherd et al. 2004). Finally, although primary producers and secondary consumers exhibited the highest mean trophic interaction strengths in the rocky-reef model, top predators exhibited by far the highest keystone indices, indicating their importance in shaping, enhancing, and perhaps stabilizing Galápagos food webs. Top predators are now rare and particularly vulnerable to fishing pressure. Continued increases in fishing pressure are likely to interact with the adverse effects associated with the intermittent and changing magnitudes of El Niño events, such as massive mortalities and severe perturbations of the food web. Understanding the combined effects of fisheries and environmental variability will thus be key to developing policies that protect the unique marine biodiversity of the Galápagos Islands.

References

Abbott, D.P. 1966. Factors influencing the zoogeographic affinities of the Galápagos inshore marine fauna. Pages 108–112 *in* R. Bowman, editor, The Galápagos. Proceedings of the symposia of the Galápagos International Scientific Project. University of California Press, Berkeley, Calif.

Beebe, W. 1924. Galápagos, world's end. Putnam, N.Y.

Belda, E.J., and A. Sánchez. 2001. Seabird mortality on longline fisheries in the western Mediterranean: factors affecting bycatch and proposed mitigating measures. Biological Conservation **98**:357–363.

Bensted-Smith, R. 2002 (editor). A biodiversity vision for the Galápagos Islands. Charles Darwin Foundation and World Wildlife Fund, Puerto Ayora, Galápagos, 147 pp. Available at www.darwinfoundation.org/articles/ft00100202.html.

Bond, W.J. 1993. Keystone species. Pages 237–291 *in* E.D. Schulze and H.A. Mooney, editors, Biodiversity and ecosystem function. Springer, N.Y.

Bow, C.S., and D.J. Geist. 1992. Geology and petrology of Floreana Island, Galápagos Archipelago, Ecuador. Journal of Volcanology and Geothermal Research **52**:83–105.

Bustamante, R.H., G.M. Wellington, G.M. Branch, G.J. Edgar, P. Martinez, F. Rivera and F. Smith. 2002a. Outstanding marine features of Galápagos. Pages 60–71 *in* R. Bensted-Smith, editor, A biodiversity vision for the Galápagos Islands. Charles Darwin Foundation and World Wildlife Fund, Puerto Ayora, Galápagos.

Bustamante, R., K.J. Collins, and R. Bensted-Smith 2000b. Biodiversity conservation in the Galápagos Marine Reserve. Bulletin de l'institute royal de Sciences naturelles de Belgique, Biologie, Supplement, **70**:31–38.

Bustamante, R.H., G.M. Branch, and R. Bensted-Smith. 2002c. The status of and threats to marine biodiversity. Pages 80–95 *in* R. Bensted-Smith, editor. A biodiversity vision for the Galápagos Islands. Charles Darwin Foundation and World Wildlife Fund, Puerto Ayora, Galápagos.

Bustamante, R.H., G.K. Reck, B.I. Ruttenberg, and J. Polovina. 2000d. The Galápagos Spiny Lobster Fishery. Pages 210–220 *in* B.F. Phillips and J. Kittaka, editors, Spiny lobsters fisheries and culture, Second edition. Fishing New Books, Blackwell Science, Oxford, U.K.

Chadwick, W.W. Jr. 2003. Bill Chadwick: Galápagos bathymetry. Available at newport.pmel.noaa.gov/~chadwick/galapagos.html.

Chavez, F.P., and R.C. Brusca 1991. The Galápagos Islands and their relation to oceanographic processes in the tropical Pacific. Pages 9–33 *in* M.J. James, editor, Galápagos marine invertebrates. Plenum Press, N.Y.

Christensen, V., and C.J. Walters. 2004. Ecopath with ecosim: methods, capabilities and limitations. Ecological Modelling **172**:109–139.

Christie, D.M., R.A. Duncan, A.R. McBirney, M.A. Richards, W.M. White, and K.S. Harpp. 1992. Drowned islands downstream from the Galápagos hotspot imply extended speciation times. Nature **355**:246–248.

Coale, K.H., K.S. Johnson, S.E. Fitzwater, S.P.G. Blaina, T.P. Stanton, and T.L. Coleya. 1998. IronEx-I, an in situ iron-enrichment experiment: experimental design, implementation and results. Deep Sea Research Part II: Topical Studies in Oceanography **45**:919–945.

Colinvaux, P.A. 1972. Climate and the Galápagos Islands. Nature **240**:17–20.

Edgar, G.J., S. Banks, J.M. Fariña, M. Calvopiña, and C. Martínez. 2004. Regional biogeography of shallow reef fish and macro-invertebrate communities in the Galápagos archipelago. Journal of Biogeography **31**:1107–1124.

Francis, M.P., L.H. Griggs, and S.J. Baird. 2001. Pelagic shark bycatch in the New Zealand tuna longline fishery. Marine and Freshwater Research **52**:165–178.

Geist, D. 1996. On the emergence and submergence of the Galápagos Islands. Noticias de Galápagos **56**:5–9.

Geist, D, T. Naumann, and P.B. Larson 1998. Evolution of Galápagos magmas: mantle and crustal level fractionation without assimilation. Journal of Petrology **39**:953–971.

Glynn, P.W. 1988. El Niño-Southern oscillation 1982–83: nearshore population, community, and ecosystem responses. Annual Review of Ecology and Systematics **19**:309–345.

Glynn, P.W. 1990. Coral mortality and disturbances to coral reefs in the tropical Eastern Pacific. Pages 55–126 *in* P.W. Glynn, editor, Global ecological consequences of the 1982–83 El-Niño-Southern Oscillation. Elsevier Oceanographic Series **52**, Amsterdam.

Glynn, P.W. 1994. State of coral reefs in the Galápagos Islands: natural vs. anthropogenic impacts. Marine Bulletin Pollution **29**:131–140.

Glynn, P.W., G.M. Wellington, and C. Birkeland 1979. Coral reefs growth in the Galápagos: limitations by sea urchin. Science **203**:47–49.

Glynn, P.W., and G. Wellington 1983. Corals and coral reefs of the Galápagos Islands. University of California Press, Berkeley.

Harpp, K.S., K.R. Wirth, and D.J. Korich 2002. Northern Galápagos Province: Hotspot-induced, near-ridge volcanism at Genovesa Island. Geology **30**:399–402.

Harris, M.P. 1969. Breeding season of sea-birds in the Galápagos Islands. Journal of Zooogy (London) **159**:145–165.

Hoff, S. 1985. Drømmen om Galápagos. Grødahl and Søn Forlag, Oslo. English translation available at www.Galápagos.to/books.htm#Hoff.

Houvenaghel, G.T. 1984. Oceanographic setting of the Galápagos Islands. Pages 43–54 *in* R. Perry, editor, Key environments: Galápagos. Pergamon Press, Oxford, U.K.

Jackson, M.H. 1994. Galápagos a natural history. Chapter 1. University of Calgary Press. Calgary, Canada.

James, M.J. 1991. Galápagos marine invertebrates—taxonomy, biogeography and evolution in Darwin's islands. Plenum Press, N.Y.

Jennings, S., A.S. Brierley, and J.W. Walker 1994. The inshore fish assemblages of the Galápagos Archipelago. Biological Conservation **70**:49–57.

Kay, E.A. 1991. The marine mollusks of the Galápagos, determinants of insular marine faunas. Pages 235–252 *in* M.J. James, editor, Galápagos marine invertebrates—taxonomy, biogeography, and evolution in Darwin's islands. Plenum Press, N.Y.

Majluf, P. 2002. Catch and bycatch of sea birds and marine mammals in the small-scale fishery of Punta San Juan, Peru. Conservation Biology **16**:1333–1343.

McCosker, J.E., and R.H. Rosenblatt 1984. Marine environment and protection. Pages 133–144 *in* R. Perry, editor, Key environments: Galápagos. Pergamon Press, Oxford, U.K.

Merlen, G. 1992. Ecuadorian whale refuge. Noticias de Galápagos **51**:23–24.

Merlen, G. 1995. A field guide to the marine mammals of Galápagos. National Institute of Fishing, Guayaquil, Ecuador.

Murillo J.C., C. Chasiluisa, B. Bautil, J. Vizcaíno, F. Nicolaides, J. Moreno, L. Molina, H. Reyes, L. García, M. Villalta, and J. Ronquillo. 2003. Pesquería de pepino de mar en Galápagos durante el año 2003. Análisis comparativo con las pesquerías 1999–2002. Pages 1–49 *in* E. Danulat, editor, Evaluación de las pesquerías en la Reserva Marina de Galápagos. Informe Compendio 2003. Fundación Charles Darwin y Dirección Parque Nacional Galápagos, Santa Cruz, Galápagos, Ecuador.

Myers, R.A., and B. Worm. 2003. Rapid worldwide depletion of predatory fish communities. Nature **423**:280–283.

Okamoto, H., and W.H. Bayliff. 2003. A review of the Japanese longline fishery for tunas and billfishes in the eastern Pacific Ocean, 1993–1997. Inter-American Tropical Tuna Commission Bulletin **22**:221–424.

Okey, T.A. 2004. Shifted community states in four marine ecosystems: some potential mechanisms. PhD Thesis, University of British Columbia, Vancouver.

Okey, T.A., S. Banks, A.R. Born, R.H. Bustamante, M. Calvopia, G.J. Edgar, E. Espinoza, J.M. Farina, L.E. Garske, G.K. Reck, S. Salazar, S. Shepherd, V. Toral-Granda, and P. Wallem. 2004. A trophic model of a Galápagos subtidal rocky reef for evaluating fisheries and conservation strategies. Ecological Modelling **172**:383–401.

Paine, R.T. 1969. A note on trophic complexity and community stability. American Naturalist **103**:91–93.

Pauly, D., V. Christensen, J. Dalsgaard, R. Froese, and F. Torres. 1998. Fishing down marine food webs. Science **279**:860–863.

Perry, R. 1984 (editor). Key environments: Galápagos. Pergamon Press, Oxford, U.K.

Power, M.E., D. Tilman, J.A. Estes, B.A. Menge, W.J. Bond, L.S. Mills, G. Daily, J.C. Castilla, J. Lubchenco, and R.T. Paine. 1996. Challenges in the quest for keystones. Bioscience **46**:609–620.

Punsly, R.G. 1983. Estimation of the number of purse seiner sets on tuna associated with dolphins in the eastern Pacific Ocean during 1959–1980. Inter-American Tropical Tuna Commission Bulletin **18**:227–299.

Ramsar, 2004. The Ramsar Convention on Wetlands. The Annotated Ramsar List: Ecuador. Available at ramsar.org/profiles_ecuador.htm.

Reck, G.K. 1983. The coastal fisheries in the Galápagos Islands, Ecuador. Description and consequences for management in the context of marine environment protection and regional development. PhD Thesis, University of Kiel.

Richmond, R.H., and P. Martinez. 1993. Sea cucumber fisheries in the Galápagos Islands: biological aspects, impacts and concerns. Technical Report submitted to the World Conservation Union (IUCN).

Roberts, G. 1999. History of world fur sealing. Available at www.parks.tas.gov.au/fahan _mi_shipwrecks/infohut/pdfs/sealingfacts.pdf.

Robinson, G., and M.E. Del Pino. 1985. El Niño in Galápagos: the 1982–1983 event. Publication of the Charles Darwin Foundation for the Galápagos Islands, Quito, Ecuador.

Ruttenberg, B. 2001. Effects of artisanal fishing on marine communities in the Galápagos Islands. Conservation Biology **15**:1691–1699.

Shepherd, S.A., P. Martinez, M.V. Toral-Granda, and G.J. Edgar. 2004. The Galápagos sea cucumber fishery: management improves as stocks decline. Environmental Conservation **31**:102–110.

Snell, H.M., P.A. Stone, and H.L. Snell. 1995. Geographical characteristics of the Galápagos Islands. Noticias Galápagos **55**:18–24.

Snell, H.M., P.A. Stone, and H.L. Snell. 1996. A summary of geographic characteristics of the Galápagos Islands. Journal of Biogeography **23**:619–624.

Tye, A., H.L. Snell, S.B. Peck, and H. Andersen 2002. Outstanding terrestrial features of the Galápagos archipelago. Pages 12–23 *in* R. Bensted-Smith, editor. A biodiversity vision for the Galápagos Islands. Charles Darwin Foundation and World Wildlife Fund, Puerto Ayora, Galápagos.

UNESCO, 1978. Intergovernmental Committee for The Protection of the World Cultural and Natural Heritage. Second Session, Washington, D.C. 5–8 September 1978. Final Report.

Uosaki, K., and W.H. Bayliff. 1999. A review of the Japanese longline fishery for tunas and billfishes in the eastern Pacific Ocean, 1988–1992. Inter-American Tropical Tuna Commission Bulletin **21**:273–488.

Vinueza, L.V., G.M. Branch, M.L. Branch, and R.H. Bustamante. 2006. Top-down herbivory and bottom-up El Niño effects on Galápagos rocky-shore communities. Ecological Monographs **76**:111–131.

Wellington, G.M. 1984. Marine environment and protection. Pages 247–263 *in* R. Perry, editor, Key environments: Galápagos. Pergamon Press, Oxford, U.K.

Whitehead, H., J. Christal, and S. Dufault. 1997. Past and distant whaling and the rapid decline of sperm whales off the Galápagos Islands. Conservation Biology **11**:1387–1396.

Wyrtki, K. 1966. Oceanography of the Eastern Equatorial Pacific Ocean. Oceanography and Marine Biology Annual Review **4**:52–68.

Wyrtki, K. 1985. Water displacements in the Pacific and the genesis of El Niño cycles, Journal of Geophysical Research **90**:7129–7132.

Zarate, P. 2002. Sharks. Pages 373–388 *in* E. Danulat and G. Edgar, editors, Ecological baseline for the Galápagos Marine Reserve. Charles Darwin Foundation, Puerto Ayora, Galápagos.

7

Food-Web Structure and Dynamics of East African Coral Reefs

Tim R. McClanahan

Physical Setting

East Africa is part of the western Indian Ocean biogeographic unit, a subregion of the world's largest biogeographic province, the Indo-Pacific. It is distinguishable from the larger province by lower species diversity than the center of diversity in South East Asia, the existence of many species widely distributed throughout the Indo-Pacific, and some regional endemism (McClanahan and Obura 1996, Sheppard 2000). East African reefs lie just south of the equator, are close to shore, and therefore are influenced by downwelling equatorial currents, tropical monsoons, and historical changes in sea levels.

Seasonal Monsoon Patterns

Despite the equatorial position of these reefs there is considerable seasonality in most of the measured physicochemical factors and this seasonality also greatly influences biotic processes (fig. 7.1). Due to the equatorial position, the Inter-tropical Convergence Zone (ITCZ) passes over the region during the two annual equinoxes. This creates two seasons of low and high wind and associated rains but the southeast monsoon, spanning April to September, is considerably wetter

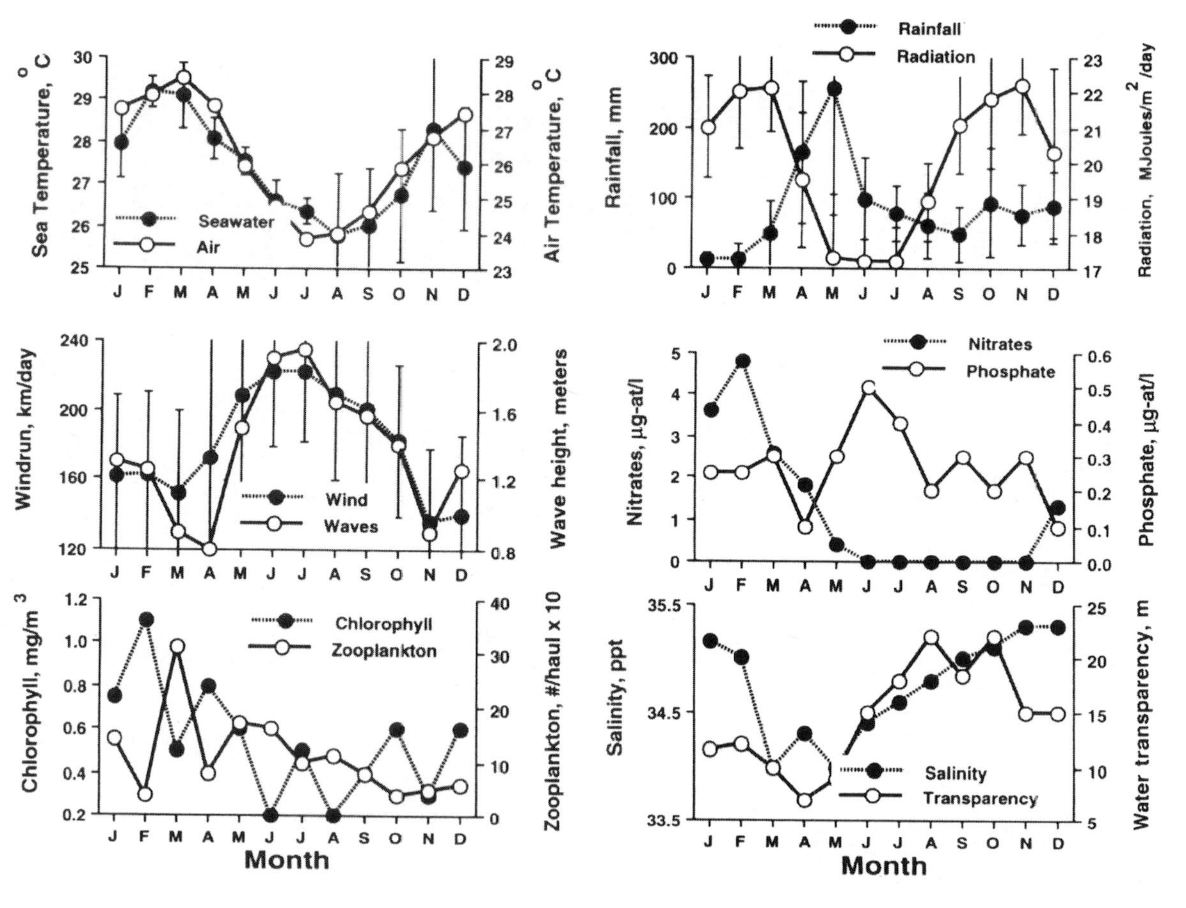

FIGURE 7.1 Seasonal changes in environmental parameters on East African reefs.

and windier than the northeast monsoon. During the southeast monsoons winds travel further across the southwestern Indian Ocean, pick up more moisture and this produces more rain, wind, water column mixing, and cooler temperatures than the northeast monsoon. The climate during the southeast monsoon is highly predictable and the cool season occurs over many months while the northeast monsoon can vary from nearly indistinguishable from background conditions to very windy and wet, but the cool and windy season rarely lasts more than two months between December and January. Peak seawater temperatures and light occur after this short cool season, in February and March and the maximum seawater temperatures will increase proportional to the weakness of the northeast monsoon.

The strength of the southeast monsoon has changed historically with a notably large increase in its intensity during the past 400 years (Anderson et al. 2002). The strength of the monsoon is positively related to northern hemisphere temperatures and fluctuations in solar insolation and, therefore, changes over glacial-interglacial periods (Fleitmann et al. 2003).

ENSO and Dipole Periodicity

The intra-annual monsoon seasons created by the latitudinal movements of the ITCZ are further influenced by two important longitudinal climatic phenomena, known as the Indian Ocean Dipole (IOD) and the El Nino-Southern Oscillation (ENSO). The IOD is caused by an intra-annual difference in heating and cooling between the eastern and western sides of the northern Indian Ocean, which may be caused by the extent of snow in Eurasia (Pang et al. 2005), and leads to propagation of warm water from the east to the west (Webster et al. 1999). The strength of the IOD varies between years and there are strong ones about every 11to 12 years (Saji et al. 1999, Cole et al. 2000).

The ENSO is similar and based in the Pacific but appears to propagate and influence climate in the Indian Ocean, with strong warm phases occurring between 3.5- and 5.3-year periods, with the shorter cycle dominant before the 1920s and after the 1960s and the longer cycle dominant in the interim (Charles et al. 1997). The East African temperature pattern is more strongly related to ENSO events than either western Australia or the Seychelles (Kuhnert et al. 1999, Cole et al. 2000), and it may be that latitudinal position, local currents, monsoonal, or local landmass effects periodically override the ENSO and oceanic forces (Charles et al. 1997). In some instances these two oscillations are in phase causing a westward propagation of unusually warm water, which was the case in 1998 when probably the largest warm-water anomaly in the past 400 years occurred (Cole et al. 2000, Anderson et al. 2002). This caused massive coral mortality in the western Indian Ocean (Goreau et al. 2000, McClanahan et al. 2007a). These intra- and interannual patterns of climate create disturbances are among the most important factors that create the foundational structure and diversity of the East African coral-reef food webs, but this foundation is greatly changed by intense resource use (McClanahan et al. 2007b).

Physico-Chemical Forces

Water flow, light, and nutrient concentrations are the most important factors influencing reef production. The strength of water flow is influenced by waves, currents, and tides and an analysis of these forces in East Africa suggest that this is also the order of their significance (McClanahan 1990). The sum of these three water-flow energy sources is estimated at 1.7×10^9 Joules/m^2/y, which is less than the estimate of solar energy of 7.1×10^9 Joules/m^2/y at the ocean surface, but still a significant contribution to reef energetics and likely to have important influences on the production of coral reefs. There is considerable seasonality in these forces as discussed above (fig. 7.1) and this seasonality also affects the concentrations of nutrients, largely through its effect on water-column mixing.

Three main forces influence phosphorus concentrations: runoff from land, upwelling, and water-column mixing. Offshore, water-column mixing is the dominant force and phosphorus concentrations are highest during the southeast monsoon when winds are strongest. Phosphorus concentrations are low to moderate for coral reefs at around 0.4 μM (tab. 7.1) but have been reported to be higher in some near-shore reef waters near towns such as Zanzibar (Bjork et al. 1995). Nitrogen is influenced by nitrogen fixation on the reef and in the water column, and runoff from land. It is highest in the open ocean during the calmest months of the year, which promotes nitrogen fixation by *Trichodesmium*, but may be high in near-shore waters due to runoff from land (Bryceson 1982, McClanahan 1988). Nitrate concentrations are 0.24 μM and total nitrogen concentrations are 3.2 μM in East African reefs. Benthic production has not been reported for this region but is likely to lie within the reported range of 1–6 gC per m^2/day for coral reefs (Kinsey 1983) and probably around 3 gC per m^2/day or around 30 g wet mass/m^2/day.

Biotic Setting

Biogeographical and Historical Setting

Taxonomically, corals are among the best-known groups in the region and studies of this group suggest that the most biologically diverse reefs in the western Indian Ocean are found among islands

Table 7.1. Description of the main trophic and functional groups found in East Africa and their estimated wet weights (kg/ha) in areas that have not been affected by heavy fishing.

Gross Trophic Level	Functional Groups	Dominant Taxa	Abundance	Reference
Primary producers	Phytoplankton	*Trichodesmium*	0.4 μg/L	Mwangi (unpubl. data)
	Seagrass	*Thalassia* and *Thallasadendron*	500 kgdw/ ha	Uku (1995)
	Algal turf	Filamentous greens and blue greens	11000 kg/ha	McClanahan (1997)
	Calcareous and coralline algae	*Halimeda Lithophyllum*, *Amphiroa* and *Jania*	3200 kg/ hand	Bjork et al. (1995)
	Frondose algae	*Turbinaria* and *Dictyota*	7780 kg/ha	McClanahan et al. (2002a)
Detritus	Plants and animals	Seagrass and algae as above	0.76%	Muzuka (2001)
Mixed primary producers/ filter feeders	Hard corals	*Porites* and *Acropora*	40% cover = 2100 kg/ ha	McClanahan et al. (2001)
	Soft corals Clams	Lobophyton *Tridacna*	4% cover <1% cover	
Detritivores	Fishes	*Actinopyga* and	<1% cover	McClanahan et al. (2001)
	Sponges	*Holothuria*		
	Sea cucumbers		150 kg/ha	Muthiga (unpubl. data)
Herbivores/ Detritivores	Zooplankton	Copepods	240 per m^3	Okemwa (1989)
	Herbivorous/ Detritivorous fish	Blennies, *Ctenochaetus*, *Chrystipera*, *Siganus* and *Scarus*	500 kg/ha	McClanahan and Kaunda-Arara (1996)
	Sea urchins	*Echinometra* and *Diadema*	10 kg/ha	
Planktivores		*Thalassoma*, *Apogon* and *Chromis*	20 kg/ha	McClanahan (1994)
Benthic invertivores	Small bodied	*Thalassoma*, *Gomphosus*, *Halichoeres*, *Centropyge* and *Chaetodon*	250 kg/ha	McClanahan and Kaunda-Arara (1996)
	Large bodied	*Coris*, *Balistapus* and *Diodon*		
Piscivores/ Invertivores		*Lutjanus*, *Lethrinus* and *Plectorhinchus*	300 kg/ha	Samoilys (1988)
Piscivores		*Epinephelus*, *Caranx* and *Carcharhinus*	100 kg/ha	Samoilys (1988)

of the central Indian Ocean, including the Maldives, Chagos, and Seychelles (Sheppard 2000). Clear water, stable salinity, and warm and stable seawater temperatures typify this region. The combination of these factors is likely to be the cause of the high diversity in this group. As one moves toward the continental margins of African and Arabia, these physicochemical conditions are less stable and diversity of corals appears to be reduced. There are 87 genera of corals in the region, of which around 50 are found in Kenya, whereas the most diverse islands of the Maldives, Chagos and Seychelles contain 55–58 genera (Sheppard 2000). Fishes may show slightly different patterns, however, as three of the better-studied groups of fish, damselfish, butterflyfish, and angelfish (Allen 1985) indicate that Sri Lanka and East Africa have the highest diversity in these groups with 92 species while the Maldives has 82 and the Seychelles 70. Consequently, the diversity of corals and fish could have different patterns in the region, depending on the environmental conditions that promote these groups.

Most of the coastline itself was created by the coral reefs during the last Pleistocene high sea level stand 130,000 years ago and these reefs have left a fossil record of coral reefs during the last interglacial period. Studies of these Pleistocene fossil reefs suggest that species distribution and diversity patterns changed over time (McClanahan 2002a). There has been a decrease in the diversity and a change in species composition of gastropod snails in East Africa since the Pleistocene. Most of the species found in the Pleistocene are still extant in the western Indian Ocean region but not found locally in East Africa, with only about 34% of the species in common between these two interglacial times. This suggests that the region is composed of meta-populations and the persistence of species is due to harlequin spatial patterns that change over time. Despite these changes in gastropod species, the fossil record suggests similar ecological organization of the dominant species with massive *Porites* forming a foundation species on which many of the branching and encrusting coral taxa colonize (Crame 1981). Crame (1986) also reports large expanses of *Acropora* beds in the Pleistocene that are not common in modern reefs.

Reef Habitats

Most of the coral reefs are fringing reefs that lie between 100 meters and a few kilometers from shore. The seaward side of the reef, or fore reef, usually drops from sea level to 10–15 meters before ending in a sand plane. The tops or reef flats are exposed during spring tides (the tidal range in the region spans 2.5–4.0 m) and the leeward back of the reef is composed of a back-reef lagoon where the percentage cover of sand and seagrass increases towards the beach. The shallow back reef is both the most studied and also most influenced by resource extraction due to the calm and shallow conditions. Most of the data and discussion that follow will be from studies of this reef habitat. Diversity within lagoons is influenced by fishing and variation in seawater temperature (McClanahan and Obura 1996, McClanahan and Maina 2003). I will first discuss the status of reefs not heavily influenced by humans followed by the current understanding of how the ecosystem has been changed by resource extraction.

The Food Web

The region has not received the same level of research on diets and energetics of reef-associated species as the Caribbean and, therefore, there are insufficient data to develop a complete food web. It is expected that the full web will be similar to the more thorough studies of the Caribbean food web (Bascompte et al. 2005). The actual species that occupy each functional group will be different, although often similar when viewed at a grosser taxonomic level of resolution such as genus or family. Given these caveats there is, nonetheless, sufficient information to determine the gross structure of the food web, its energetics and some of the more important interactions. Below is a brief description of the food web in situations largely uninfluenced by human resource extraction.

The main sources of benthic primary production are the various forms of algae that live on dead coral, the symbiotic algae living in coral, and seagrass (tab. 7.1). Given the estimated biomass and probable production values stated above, it is quite likely that the wet weight of this production is turned over on the order of every 50 days, but considerably less for the microscopic algae that have turnovers of around 20–30 days (McClanahan 1997) and longer for large frondose algae and seagrass. Regardless, it indicates that much of the consumers' wet weight is maintained by a rapid turnover of primary producers, with production going quickly into the grazing and detrital parts of the food web.

The amount of production that goes into herbivores versus detritus and detritivores is not currently known as it has recently been recognized that many "grazing" fishes are, in fact, feeding on detritus (Choat et al. 2004). The combined biomass of herbivores and detritivores is high and around 500 kg/ha (McClanahan et al. 2007b), but how much of this biomass depends on these two sources of production has yet to be determined. To maintain this biomass and assuming little loss of energy in the conversion of primary production to detritus, these grazing fish would consume an equivalent of 6% of their body weight per day in benthic production. Careful studies of parrotfish suggest that they consume an equivalent of around 10% or more of their body weight per day (Bruggemann et al. 1994). There are, therefore, some unknown sources of production required to explain the balance of primary production and consumption. The difference may be due to the unknown value of benthic productivity for East African reefs, the production and partitioning of detritus, algae, and plants in their diet (Choat et al. 2004), and possibly some import of production or concentration of herbivore/detritivores on the coral reefs. Given that many coral reefs in East Africa are surrounded by seagrass meadows, it may be that this seagrass production is exported to the coral reef and subsidizing detrital grazers on the coral reef. The detrital pathway is difficult to measure and the organic fraction of the sediments is small at only 0.76%, which suggests that detritus may be like primary producers in having a low mass but high turnover.

The common herbivores/detritivores include damselfishes, surgeonfishes, parrotfishes, rabbitfishes, blennies, and rudderfishes, with surgeonfishes and parrotfishes each having a wet weight of about 250 kg/ha and herbivorous damselfishes being very common but probably contributing less weight at around 50 kg/ha. Rudderfishes are patchily distributed, which makes estimate of their wet weights difficult, but their mass is probably not large. Invertebrate herbivores may be quite common in intertidal reefs but, in unfished subtidal reefs being discussed here, they are not common, with the most common invertebrate grazers, sea urchins, estimated at 10 kg/ha.

The wet weight of phytoplankton in the water is at least three orders of magnitude lower than the wet weight of benthic algae, and this results in a low wet weight estimate of 20 kg/ha for planktivores on these reefs. Consequently, this trophic pathway is not large, but is expected to vary spatially with water movement and potentially has a high turnover. The common species of planktivores include species such as fusiliers, damselfishes in the genera *Chromis, Neopomacentrus* and *Dascyllus*, and anthiases. Fusiliers and anthiases are only common on the fore reef, whereas damselfishes are relatively more abundant in the lagoon.

The third trophic level is a mixture of fish that eat various invertebrates and fish. There is a very high diversity of species at this level of the food web with more than half of the 256 common species found in fisheries catches and ecological surveys positioned at this trophic level (fig. 7.2). Despite the high diversity, the total estimated wet weight of these two functional groups, that include primarily benthic invertivores and piscivores/invertivores, is only around 550 kg/ha. So, whereas about 49 species of fish are primarily herbivores/detritivores at tropic level 2.0–2.5 there are 123 and 56 species that occupy trophic levels 2.5–3.5 and 3.5–4.0, respectively. This suggests high ecological complexity at this position in the food web and the maintenance of high diversity with relatively low food and energy resources.

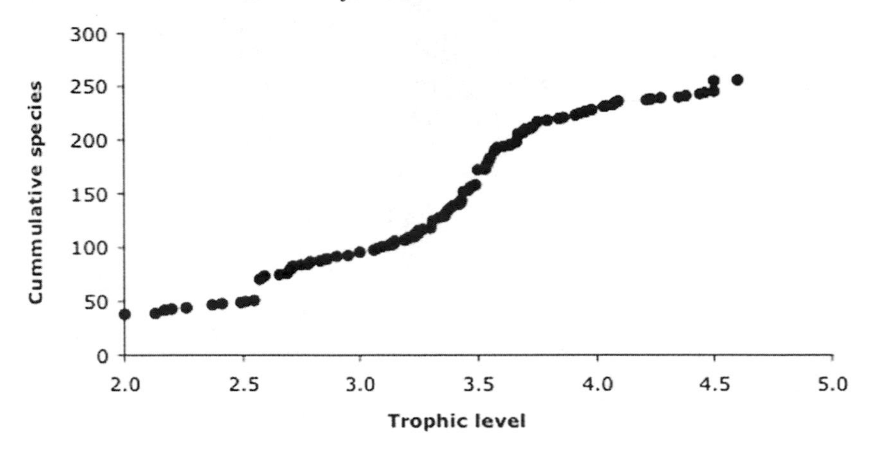

FIGURE 7.2 Cumulative frequency plot of the common 256 fish species found on Kenyan reefs as a function of their mean trophic level. The fish species list is based on the fish studied in belt transects (McClanahan 1994) and caught at fish landing sites (McClanahan and Mangi 2001). Trophic level data are from Froese and Pauly (2000).

Common invertivores include those that feed primarily on small, free-living, and cryptic invertebrates, often various types of crustaceans and worms, but also juveniles of other groups. These are largely the small-bodied wrasses and smaller individuals of triggerfish, emperors, butterflyfish, and snappers. Many of the butterflyfish are coral specialists but some have broader diets. Angelfish primarily feed on sponges, and the larger triggerfish, emperors, and wrasses will feed on the larger invertebrates (McClanahan 1995). Octopus and lobster would be among the few invertebrates that occupy this higher trophic level.

Only 28 species are common above trophic level four and these include piscivores such as barracuda, cod, jacks, trumpetfish, and sharks. Some invertebrates such as a few species of cone shells also feed primarily on fish, but their abundance and metabolism are low and expected to be ecologically unimportant (McClanahan 2002c). This higher trophic level is probably the most poorly known, due to the difficulty of studying species that exist at low population densities (Samoilys 1988).

Influence of Resource Extraction

The above description of the food web is largely based on studies in marine protected areas but studies of the reefs with heavy resource extraction have shown that the food web is altered considerably by this extraction (McClanahan et al. 1999). These changes will be described first with a simple simulation model and then by comparing the model simulation with field studies of reefs experiencing different levels and changes in resource extraction.

An aggregated food web A highly simplified food web and associated simulation model was developed for the purpose of determining the influence of fishing intensity and choices on the ecosystem (McClanahan 1995, fig. 7.3). As described above, the planktonic portions of the food web are not large and by combining the detritus and primary production pathways it is possible to reduce the complexity of the food web and make it more tractable for simulation modelling. The model estimates benthic primary production and its passage through consumers and fishermen. The model also estimates inorganic carbon or calcium-carbonate deposition by corals and algae, and erosion by herbivorous fish and grazing sea urchins into sand. As described above many species are pooled into the larger functional groups of the model but the wet weights of the groups are similar to those presented in table 7.1 and consumption and production rates of these groups are estimated from various coral reef studies. Catch rates are also estimated for artisanal fisheries, where the maximum catch per fishermen is estimated at 25 kg/ha/day in reefs with no prior fishing.

During model simulations the number of fishermen and their prey selection can be changed by the computer program to determine the outcome on the harvesting rate of each group on the rest of the ecosystem and the organic and inorganic carbon. For example, fishermen may select piscivores, invertivores or herbivores/detritivores or any of the possible combinations, the model can be run until no change or a steady state is achieved (39 years), and then the variables in the model are plotted against the density of fishermen.

A few of the key aspects of model calibration that increase understanding of the model are: (a) that algae and corals differ in that corals have high inorganic but low organic production compared to algae; (b) sea urchins have low consumption and production but high biomass compared to grazing fish; and (c) this results in high rates of reef erosion by urchins compared to herbivorous fish. Consequently, as the ratios of these primary producers and consumer's change, it has important ecological consequences for the carbon-based processes of the reef.

Model simulations Results of this simple model indicate the complexity of interactions and the ecological effects of fishing. Fishing all groups (fig. 7.4a) leads to the elimination of piscivores and invertivores at 0.03 and 0.09 fishermen/ha respectively. A large change occurs, however, at 0.07 fishermen/ha where the maximum catch of around 300 kg/ha/year is reached. Here, sea urchin wet weight increases to 4500 kg/ha and this reduces algal wet weight, gross and

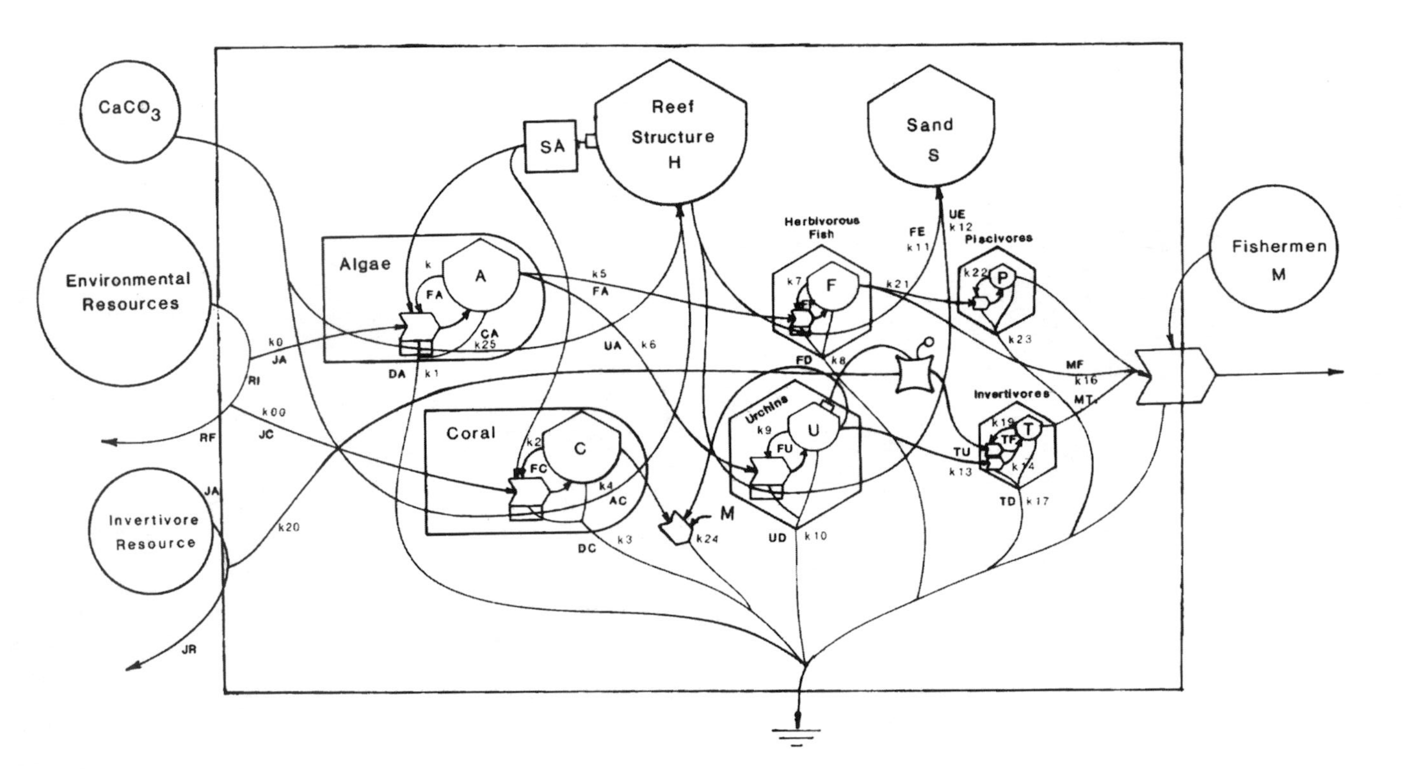

FIGURE 7.3 An energy-flow diagram of the East African coral reefs used to test for the effects of the intensity and choices of the fishing on the ecosystem. The system is contained within a box and circles are energy or material flows from outside of the system. Lines are flows of energy or materials, arrows are interactions, tanks are storages, primary producers are bullets, and consumers are hexagons. The full details of the model, symbols, abbreviations and calculations are given in McClanahan (1995).

FIGURE 7.4 Simulation outputs for a coral reef where (a) all groups; (b) only piscivores; and (c) piscivores and herbivores are fished. For each fishing regime, the changes in algae, coral, urchins, and fish (left column) as well as organic and inorganic carbon production and fisheries yields (right column) are shown in relation to fishing intensity. GPP, gross organic carbon productivity; NPP, net organic carbon productivity; CPUE, catch per unit effort. Hard substratum is the weight while growth is the daily production of calcium carbonate. Modified from McClanahan (1995).

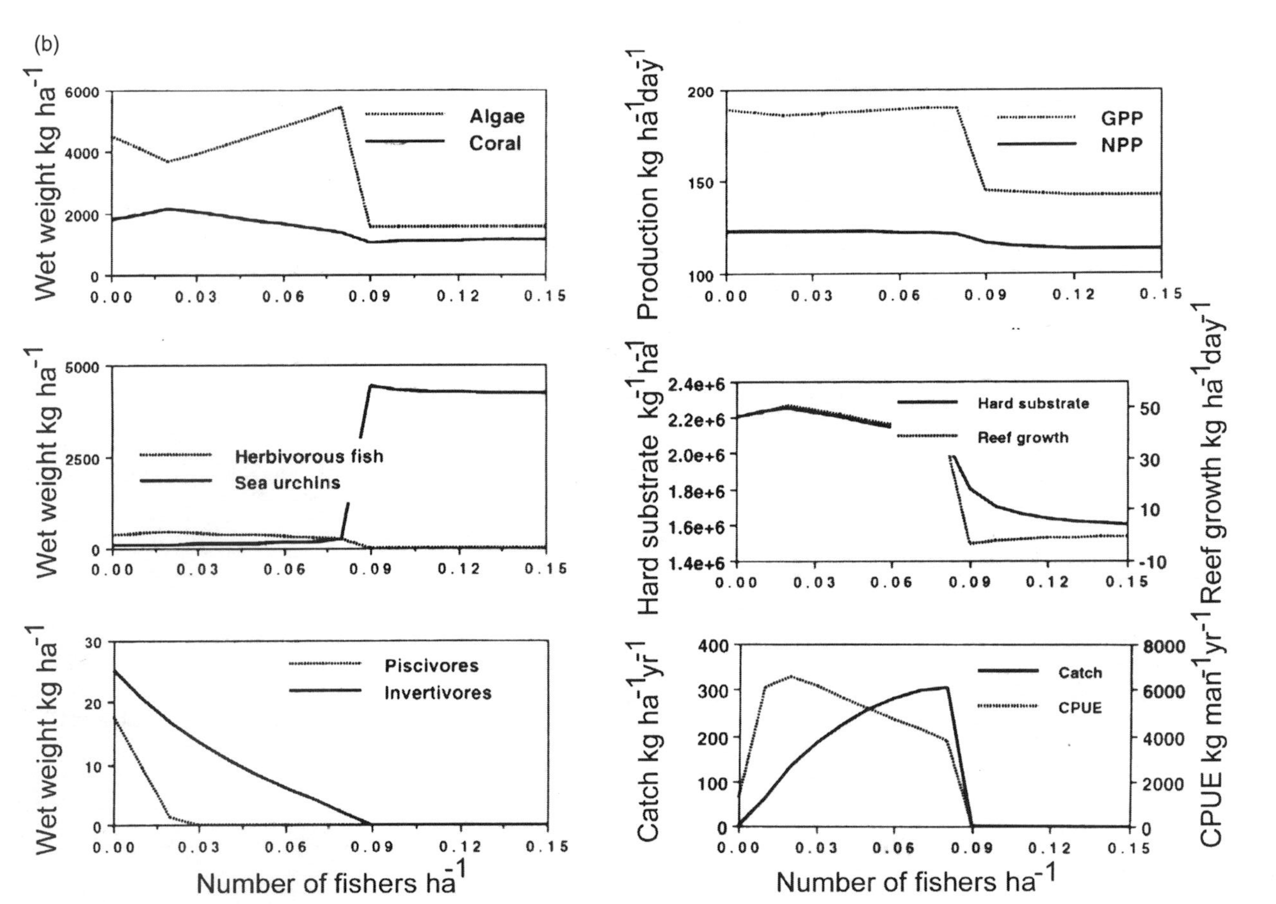

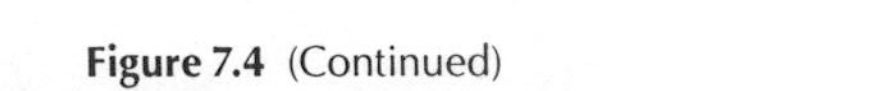

Figure 7.4 (Continued)

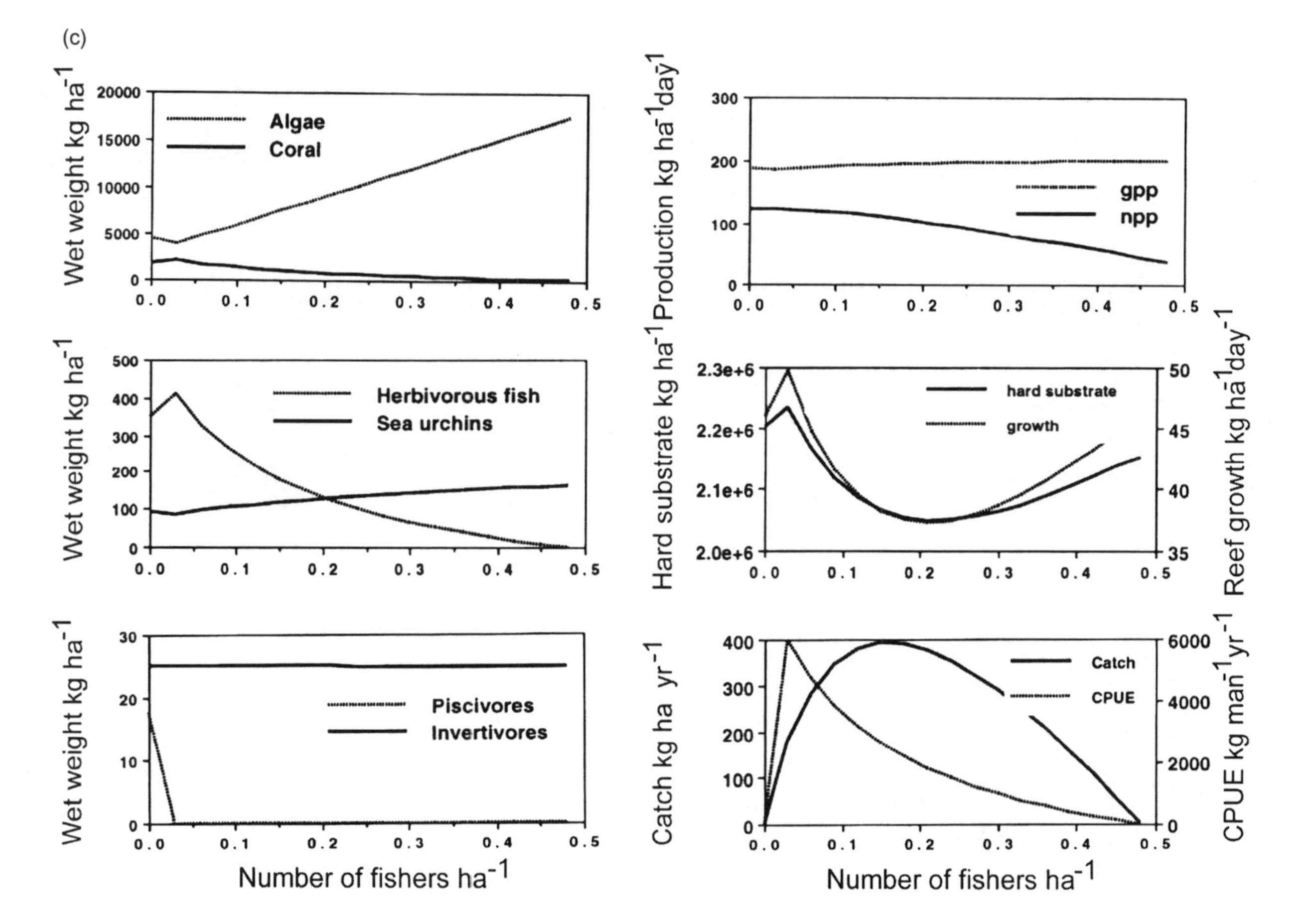

Figure 7.4 (Continued)

net organic carbon productivity. Inorganic carbon or reef growth is also reduced from 50 to 8 kg/ha/day.

The rapid change and loss of fisheries catches and reef process suggest the need for alternatives and possibly the need to focus fishing on piscivores and herbivores in order to maintain invertivores at a level that will not allow sea urchins to dominate. Just fishing piscivores (fig. 7.4b) has the effect of keeping sea urchin at low levels and also results in an increase in herbivorous fish, a reduction in algae, and an increase in coral and reef growth to 57 kg/ha/day. This would appear to be a benign form of fishing for the ecosystem, but the maximum yield of fish is less than 3 kg/ha/year and achieved at 0.02 fishermen/ha. This is not likely to be effective in areas near large human populations.

Fishing piscivores and herbivorous fish only (fig. 7.4c) also keeps sea urchins from dominating, and the yields are high at 400 kg/ha/year at 10 times the fishing effort, or 0.2 fishermen/ha. The ecological effect of this fishing regime is, however, less benign as algal wet weight doubles and net productivity is reduced where the maximum yield is achieved. This also has the effect of reducing coral and reef growth to 37 kg/ha/day. Somewhat unexpectedly the model predicts an eventual increase in reef growth above this maximum yield and this is due to calcification by algae alone, with little contribution from corals. This will, of course, depend on the accuracy of the model's estimate of calcification by algae, which will depend greatly on the abundance and success of calcifying algal species in these algal-dominated communities.

Comparisons with field data The above model provides insights into possible ways to manage the reef and outcomes of different fishing scenarios. Of course it suffers the problems of being highly simplified in that the approximately 250 species of fish that occupy most coral reefs and their complex diets are pooled into only three functional groups. It behoves us to know if this species complexity is or is not important for the gross level of interaction, ecological processes, and fisheries yield that most influence decisions about fisheries regulations. To further test the model's relevance and predictive ability we can compare model results with field studies of fished, unfished, and reefs in transition between these two states.

Densities of fishers at studied Kenyan reefs are estimated to exceed 0.07 per ha and therefore fishing is above the level where the predicted switch to sea urchin dominance is predicted (McClanahan and Mangi 2001). These reefs have sea urchin wet weights of around 4000 kg/ha, one-half the coral cover of unfished reefs, and very low wet weights of all other fish groups, with estimates for all fish being <250 kg/ha compared to values between 1000–1500 kg/ha in marine protected areas (McClanahan and Graham 2005). These fished reefs still produce fish, but the dominant fisheries species are the seagrass-feeding rabbitfish and parrotfish and other sand and seagrass-associated generalist and planktonic species that are not highly dependent on the hard-bottom coral-reef environment.

Comparisons of model steady states with fished and unfished reefs in Kenya and Tanzania suggest similarities that are not due to calibration of the model with East African field data (McClanahan 1995, McClanahan et al. 1999). Data used to calibrate the model are from coral-reef studies from around the tropics and the equations used do not force a steady state for the various variables at predetermined levels. Rather, the model structure depends most on good estimates of minimum resource level and consumption and metabolism of the functional groups (McClanahan 1992), and the steady states only arise from the interaction of the functional groups and their production and consumption rates. Consequently, the good fit between model steady states for fished and unfished reefs and field data suggest reasonable estimates of the energetic processes.

A reef in transition from heavy to no fishing in Mombasa around 1990 provides a further test of the model. Comparing the gross functional groups over time indicates increases in the invertebrate predators and their predation rates and decreases in sea urchin wet weights (fig. 7.5). There were also increases in coral cover and herbivorous fish, although much of the coral cover died during 1998 due to unusually warm water in that year. A study of the important sea urchin predators over time showed a succession in the species that eat the urchins (McClanahan 1995). During the early stages after park designation, the triple-lobed wrasse was the dominant predator but it was eventually replaced by the red-lined triggerfish, which is the dominant predator in older parks.

Thirteen years after the park was designated in Mombasa there are still some groups that might be expected to change further. For instance, although sea urchin wet weights were reduced from 6,000 to 2,000 kg/ha over this period, the model would suggest that they should be reduced further. The levelling at 2,000 kg/ha may be due to poor recovery of some of the specialized sea urchin predators, or because the dominant sea urchin in the park is the largest local species, *Echinothrix diadema* (≈350 g), which it is more immune to predation than other species used in the calibration. Herbivorous parrotfish have recovered well, but recovery has been slower for the surgeonfish (McClanahan et al. 2007b). These observations suggest that model predictions based on energetics may be faster than actual values. This may be due to important life-history characteristics of species that are not well accounted for when pooling these groups into generic functional groups in the model or possibly the edge and spillover effect created by small parks (McClanahan and Mangi 2000). Nevertheless, the model does have some predictive

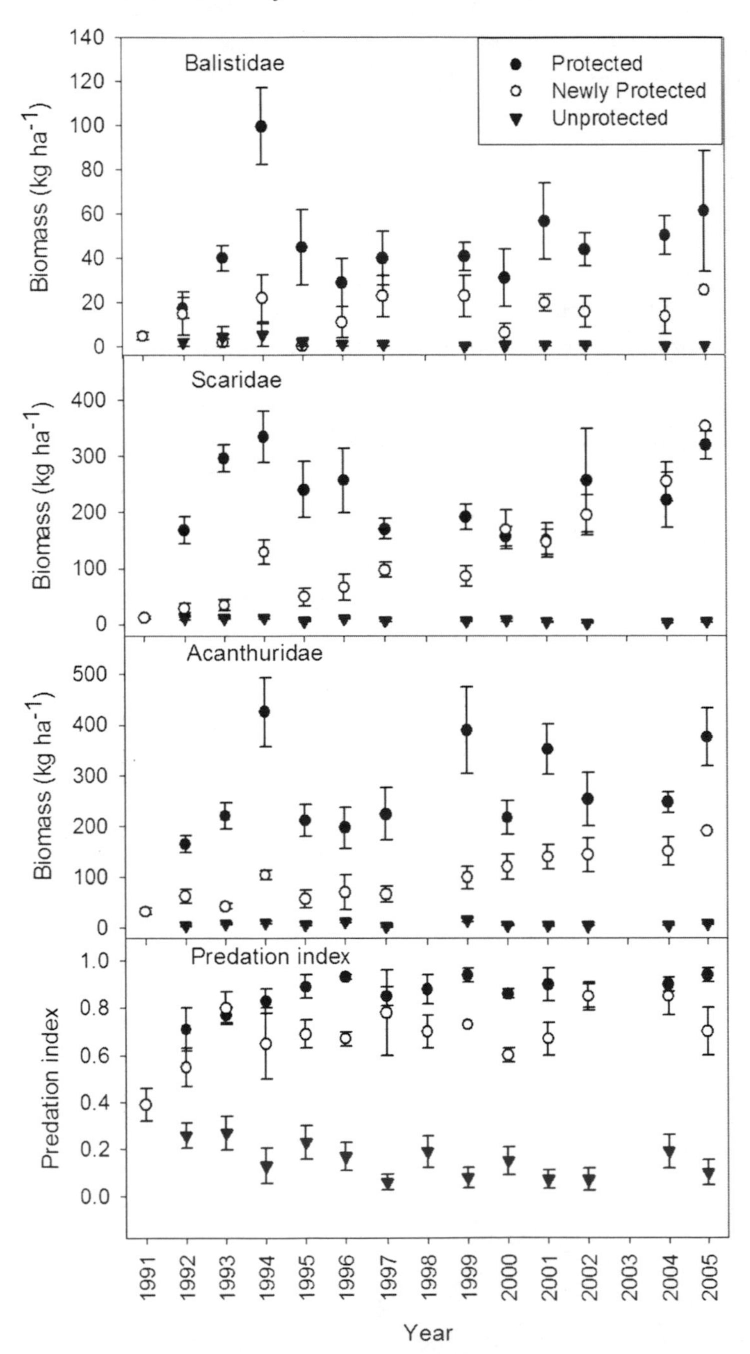

FIGURE 7.5 Plots of the change in the invertivorous triggerfish (Balistidae), herbivorous/detritivorous parrotfish (Scaridae) and surgeonfish (Acanthuridae) and the relative predation rates on tethered sea urchins *Echinometra mathaei* in two old protected parks (Malindi and Watamu MNP), a newly protected park (Mombasa MNP), and four unprotected and heavily fished reefs (Diani, Raslwatine, Kanamai, and Vipingo). Mombasa was converted to a marine park in 1991.

ability and will form the basis for further, more sophisticated, models as more information on the parameters and species' life histories accumulates.

Climatic Influences

Intra-annual patterns As described above, coral reefs in this region are exposed to both intra-annual seasonality and interannual climatic oscillations. Seasonality is relatively well studied and understood (McClanahan 1988) but interannual patterns have only recently received attention and appreciation. Seasonal patterns greatly influence reproduction and recruitment patterns, but the effect on populations of the few species studied indicates that high mortality in the post-settlement stages dampens potential oscillations that these seasonal patterns might produce in adult populations (Muthiga 1996). Consequently, most studies of large invertebrates and fish assemblages across seasons do not produce notable changes, even when potential resources such as benthic algal cover change seasonally (McClanahan et al. 2002a).

Benthic algae appear most affected by seasonal patterns with turf and frondose algae exhibiting the opposite patterns (McClanahan 1997, McClanahan et al. 2002a). Turfs reach their highest wet weights during the cool southeast monsoon and frondose algae during the warm northeast monsoon. The causes of these patterns have not been experimentally determined but it is likely that turf algae do well in the cool season due to the lower water temperatures, which are expected to reduce grazing pressure (Polunin and Klump 1989), and through some assistance from higher seawater phosphorus concentrations during this season. The build up of frondose algae, on the other hand, may be more associated with the calmer physical conditions during the warm season. It is notable that as the seas change from calm to rough, many of these large algae are ripped up and maintained at low levels until the monsoon switches again. Experimental studies with nutrients and grazing fishes in other regions show that turfs are more positively influenced than frondose algae by the addition of nutrients, and grazing has the opposite effects (McClanahan et al. 2004).

Interannual patterns Periodic coral bleaching and mortality events have increased attention on interannual patterns. These events have affected this region, often on the scale of the warm ENSO and IOD events, reoccurring every 3.5 to 5.3 years (Huppert and Stone 1998). Since first recorded in this region in 1987, bleaching events have been observed in 1994, 1998, 2003, and 2005. The 1987 and 1998 events caused significant mortality of hard and soft coral while these groups survived considerably better in the other bleaching years. Consequently, there may be an approximately decadal cycle of mortality and associated

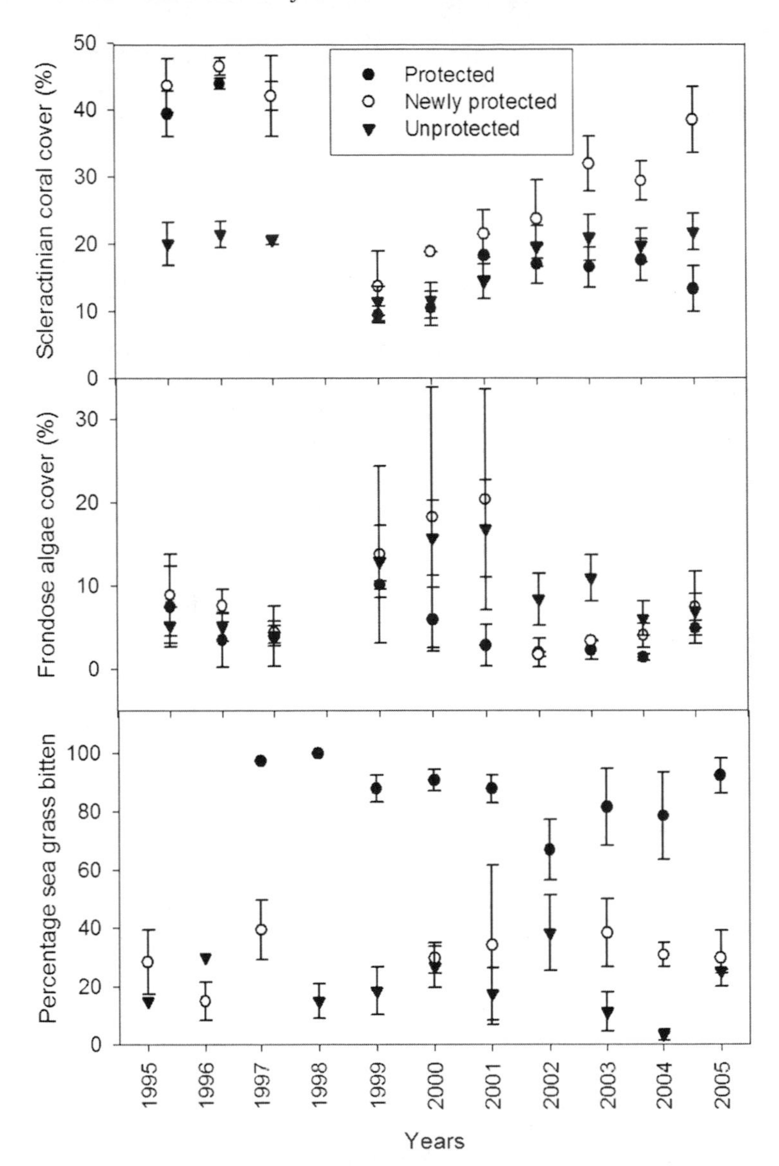

FIGURE 7.6 Changes in benthic cover and herbivory with time across the 1998 coral bleaching event. Changes in scleractinian coral (top), frondose algal cover (middle), and herbivory rates (bottom) as determined by the percentage of the seagrass *Thalassia hemprichii* bitten in 24 hours in the three management categories described in figure 7.5. Bars are standard errors of the mean based on sites.

recovery for these symbiotic groups and this is expected to have consequences for other components of the food web.

Monitoring studies indicate some resilience to these events and strong interaction between fishing and climatic disturbances (fig. 7.6). During the early 1990s coral cover in unfished marine protected areas (MPAs) was two times higher than fished reefs and this resulted in considerably higher loss of coral cover during the 1998 ENSO. Recovery rates of coral cover after 1998 were, however, similar for both management systems after 5 years, with almost complete recovery to the pre-1998 composition in fished reefs but recovery of only half of the pre-1998 level in marine parks. This is attributable to the higher initial levels and slow recovery of sensitive branching coral taxa in the unfished marine parks compared to fished reefs (McClanahan et al. 2001, McClanahan and Maina 2003). Coral diversity was depressed after the event but recovered to pre-1998 levels after 3–5 years. Total fisheries catches were largely unaffected by the event despite some small changes in the abundance of fish (McClanahan et al. 2002b).

Frondose algal cover increased on all reefs across the 1998 ENSO event. It was, however, reduced to its original level after 4 years in unfished reefs while remaining high in fished reefs. Diversity of the frondose algae was high and increased in fished reefs, while lower and unchanged in the unfished reefs. This is attributable to the higher grazing rates by fishes in the parks, which increased across this 1998 event in unfished but not fished reefs.

Summary

East African coral reefs are of intermediate to high diversity on a global scale and experience moderate levels of seasonal and interannual disturbances. Studies of water temperature and the intensity of the monsoons suggest that climate in this region has been warming for the past four centuries, with most of it in the last 50 years (Cole et al. 2000), and this is associated with an increasing intensity of monsoons, ENSO, and IOD disturbances. There is a concern that these cycles will continue to occur on a rising baseline of seawater temperatures, and that the frequency of these events will increase beyond what will allow sufficient time for recovery (Hughes et al. 2003, Sheppard 2003). Consequently, this increasing trend, in conjunction with human resource extraction pressures, will lead to an increase in the level of disturbance and the number of reefs in states of recovery. This creates significant challenges for management (McClanahan 2002b, Hughes et al. 2003). This study and others suggest the importance of maintaining populations of herbivores to prevent algal wet weight from increasing and competing with corals. In East African back reefs this is maintained by fish in the parks and sea urchin populations in the fished reefs but with different consequences for reef diversity and ecological processes. Most studies to date have focused on the back-reef habitat. Therefore, to improve understanding of climate and resource use changes, there is a need to study the consequences of interactions in different reef habitats.

References

Allen, G.R. 1985. Butterfly and angelfishes of the world. Mergus Publishers, Melle, Germany.

Anderson, D.M., J.T. Overpeck, and A.K. Gupta. 2002. Increase in the Asian Southwest Monsoon during the past four centuries. Science **297**:596–599.

Bascompte, J., C. Melian, and E. Sala. 2005. Interaction strength combinations and the overfishing of a marine food web. Proceedings of the National Academy of Science USA **102**:5443–5447.

Bjork, M., S.M. Mohammed, M. Bjorklund, and A. Semesi. 1995. Coralline algae, important coral-reef builders threatened by pollution. Ambio **24**:502–505.

Bruggemann, J.H., J. Begeman, E.M. Bosma, P. Verburg, and A.M. Breeman. 1994. Foraging by the stoplight parrotfish *Sparisoma viride*. II. Intake and assimilation of food, protein and energy. Marine Ecology Progress Series **106**:57–71.

Bryceson, I. 1982. Seasonality of oceanographic conditions and phytoplankton in Dar es Salaam waters. University Science Journal (Dar University) **8**:66–76.

Charles, C. D., D.E. Hunter, and R.D. Fairbanks. 1997. Interaction between the ENSO and the Asian Monsoon in a coral record of tropical climate. Science **277**:925–928.

Choat, J.H., W.D. Robbins, and K.D. Clements. 2004. The trophic status of herbivorous fishes on coral reefs. Marine Biology **145**:445–454.

Cole, J., R. Dunbar, T. McClanahan, and N. Muthiga. 2000. Tropical Pacific forcing of decadal variability in SST in the western Indian Ocean. Science **287**:617–619.

Crame, J.A. 1981. Ecological stratification in the Pleistocene coral reefs of the Kenya Coast. Palaeontology **24**:609–646.

Crame, J.A. 1986. Late Pleistocene molluscan assemblages from the coral reefs of the Kenya coast. Coral Reefs **4**:183–196.

Fleitmann, D., S.J. Burns, M. Mudelsee, U. Neff, J. Kramers, A. Mangini, and A. Matter. 2003. Holocene forcing of the Indian Monsoon recorded in a stalagmite from southern Oman. Science **300**:1737–1739.

Froese, R., and D. Pauly, editors. 2000. FishBases 2000: concepts, design and data source. ICLARM, Makati City, Philippines.

Goreau, T., T. McClanahan, R. Hayes, and A. Strong. 2000. Conservation of coral reefs after the 1998 global bleaching event. Conservation Biology **14**:5–15.

Hughes, T.P., A.H. Baird, D.R. Bellwood, M. Card, S.R. Connolly, C. Folke, R. Grosberg, O. Hoegh-Guldberg, J.B.C. Jackson, J. Kleypas, J.M. Lough, P. Marshall, M. Nystrom, S. Palumbi, J.M. Pandolfi, B. Rosen, and J. Rougharden. 2003. Climate change, human impacts, and the resilience of coral reefs. Science **301**:929–933.

Huppert, A., and L. Stone. 1998. Chaos in the Pacific's coral bleaching cycle. American Naturalist **152**:447–459.

Kinsey, D.W. 1983. Standards of performance in coral reef primary production and carbon turnover. Pages 209–220 *in* D. J. Barnes, editor. Perspectives on coral reefs. Brian Clouston Publisher, Manuka.

Kuhnert, H., J. Patzold, B.G. Hatcher, K.-H. Wyrwoll, A. Eisenhauer, L.B. Collins, Z.R. Zhu, and G. Wefer. 1999. A 200-year coral stable oxygen

isotope record from a high-latitude reef off Western Australia. Coral Reefs **18**:1–12.
McClanahan, T.R. 1988. Seasonality in East Africa's coastal waters. Marine Ecology Progress Series **44**:191–199.
McClanahan, T.R. 1990. Hierarchical control of coral reef ecosystems. PhD dissertation. University of Florida, Gainesville.
McClanahan, T.R. 1992. Resource utilization, competition and predation: a model and example from coral reef grazers. Ecological Modelling **61**:195–215.
McClanahan, T.R. 1994. Kenyan coral reef lagoon fish: effects of fishing, substrate complexity, and sea urchins. Coral Reefs **13**:231–241.
McClanahan, T.R. 1995. A coral reef ecosystem-fisheries model: impacts of fishing intensity and catch selection on reef structure and processes. Ecological Modelling **80**:1–19.
McClanahan, T.R. 1997. Primary succession of coral-reef algae: differing patterns on fished versus unfished reefs. Journal of Experimental Marine Biology and Ecology **218**:77–102.
McClanahan, T.R. 2002a. A comparison of the ecology of shallow subtidal gastropods between western Indian Ocean and Caribbean coral reefs. Coral Reefs **21**:399–406.
McClanahan, T.R. 2002b. The near future of coral reefs. Environmental Conservation **29**:460–483.
McClanahan, T.R. 2002c. The effects of time, habitat and fisheries management on Kenyan coral-reef associated gastropods. Ecological Application **12**:1484–1495.
McClanahan, T. R., M. Ateweberhan, C. R. Sebastian, N. A. J. Graham, S. K. Wilson, M. M. M. Guillaume, and J. H. Bruggemann. 2007a. Western Indian Ocean coral communities: bleaching responses and susceptibility to extinction. Marine Ecology Progress Series **337**:1–13.
McClanahan, T.R., and N.A.J. Graham. 2005. Recovery trajectories of coral reef fish assemblages within Kenyan marine protected areas. Marine Ecology Progress Series **294**:241–248.
McClanahan, T. R., N. A. J. Graham, J. M. Calnan, and M. A. MacNeil. 2007b. Towards pristine biomass: reef fish recovery in coral reef marine protected areas in Kenya. Ecological Applications **17**:1055–1067.
McClanahan, T.R., and B. Kaunda-Arara. 1996. Fishery recovery in a coral-reef marine park and its effect on the adjacent fishery. Conservation Biology **10**:1187–1199.
McClanahan, T.R., and J. Maina. 2003. Response of coral assemblages to the interaction between natural temperature variation and rare warm-water events. Ecosystems **6**:551–563.
McClanahan, T.R., J. Maina, and L. Pet-Soede. 2002b. Effects of the 1998 coral mortality event on Kenyan coral reefs and fisheries. Ambio **31**:543–550.
McClanahan, T.R., and S. Mangi. 2000. Spillover of exploitable fishes from a marine park and its effect on the adjacent fishery. Ecological Applications **10**:1792–1805.
McClanahan, T.R., and S. Mangi. 2001. The effect of closed area and beach seine exclusion on coral reef fish catches. Fisheries Management and Ecology **8**:107–121.

McClanahan, T.R., N.A. Muthiga, A.T. Kamukuru, H. Machano, and R. Kiambo. 1999. The effects of marine parks and fishing on the coral reefs of northern Tanzania. Biological Conservation **89**:161–182.

McClanahan, T.R., N.A. Muthiga, and S. Mangi. 2001. Coral and algal response to the 1998 coral bleaching and mortality: interaction with reef management and herbivores on Kenyan reefs. Coral Reefs **19**:380–391.

McClanahan, T.R., and D.O. Obura 1996. Coral reefs and nearshore fisheries. Pages 67–99 *in* T. R. McClanahan and T. P. Young, editors. East African ecosystems and their conservation. Oxford University Press, N.Y.

McClanahan, T.R., E. Sala, P.J. Mumby, and S. Jones. 2004. Phosphorus and nitrogen enrichment do not enhance brown frondose "macroalgae." Marine Pollution Bulletin **48**:196–199.

McClanahan, T.R., J.N. Uku, and H. Machano. 2002a. Effect of macroalgal reduction on coral-reef fish in the Watamu Marine National park, Kenya. Marine and Freshwater Research **53**:223–231.

Muthiga, N.A. 1996. The role of early life history strategies on the population dynamics of the sea urchin *Echinometra mathaei* (de Blainville) on reefs in Kenya. PhD dissertation, University of Nairobi, Kenya.

Muzuka, A.N.N. 2001. Sources of organic matter in the Msasani Bay and Dar es Salaam Harbour. Pages 61–76 *in* M.D. Richmond and J. Francis, editors. Marine science development in Tanzania and Eastern Africa. WIOMSA, Zanzibar, Tanzania.

Okemwa, E.N. 1989. Analysis of six 24-hour series of zooplankton samplings across a tropical creek, the Port Reitz, Mombasa, Kenya. Tropical Zoology **2**:123–138.

Pang, H., Y. He, A. Lu, J. Zhao, B. Song, B. Ning, and L. Yuan. 2005. Influence of Eurasian snow cover in spring on the Indian Ocean Dipole. Climate Research **30**:13–19.

Polunin, N.V.C., and D.W. Klump. 1989. Ecological correlates of foraging periodicity in herbivorous reef fishes of the Coral Sea. Journal of Experimental Marine Biology and Ecology **126**:1–20.

Saji, N.H., B.N. Goswami, P.N. Vinayachandran, and T. Yamagata. 1999. A dipole mode in the tropical Indian Ocean. Nature **401**:360–363.

Samoilys, M.A. 1988. Abundance and species richness of coral reef fish on the Kenyan Coast: The effects of protective management and fishing. Proceedings of the Sixth International Coral Reef Symposium **2**:261–266.

Sheppard, C.R.C. 2000. Coral reefs of the western Indian Ocean: An overview. Pages 3–38 in T.R. McClanahan, C.R.C. Sheppard, and D.O. Obura, editors. Coral reefs of the Indian Ocean: their ecology and conservation. Oxford University Press, N.Y.

Sheppard, C.R.C. 2003. Predicted recurrences of mass coral mortality in the Indian Ocean. Nature **425**:294–297.

Uku, J.N. 1995. An ecological assessment of littoral seagrass communities in Diani and Galu coastal beaches, Kenya. Msc dissertation (Biology of Conservation), University of Nairobi, Kenya.

Webster, P.J., A.M. Moore, J.P. Loschnigg, and R.R. Leben. 1999. Coupled ocean-atmosphere dynamics in the Indian Ocean during 1997–98. Nature **401**:356–359.

8

Food-Web Structure and Dynamics of Eastern Tropical Pacific Coral Reefs: Panamá and Galápagos Islands

Peter W. Glynn

Studies of the food-web structure and predator–prey interactions of coral reefs, among the most structurally complex and species rich marine ecosystems known, reveal a plethora of diverse consumer links and dynamics (Glynn 1988a, Opitz 1996, Pennings 1997, McClanahan 2004). This is true even for the low-diversity coral reef off the Pacific coast of Panamá, in the most remote and species-poor biogeographic province of the Indo-Pacific region. A long-term study of the coral reef at Uva Island, Panamá has revealed 16 guilds and 31 inter-guild links in a generalized food web, and a coral–corallivore subweb composed of eight invertebrate and 12 fish corallivore species with no less than 287 species links (Glynn 2004). This degree of complexity in a low-diversity ecosystem rivals that reported for coral reefs in Hawaii, Okinawa, Marshall Islands, and on the Great Barrier Reef (Karlson 1999).

In this chapter, the Uva Island coral-reef food web is examined in terms of (1) a topological or descriptive portrayal of feeding links, and (2) interaction relationships of consumer feeding links that can regulate coral community species composition and structure. I will also present a subweb of cryptic species that reside in coral reef frameworks. Intense predation within reef framework structures often goes unnoticed in coral reef food-web analyses. A final topic addresses the severe effects of El Niño disturbances and how these have influenced the structure of coral-reef food webs in the Galápagos Islands.

The Environment

Physical Setting

Some of the broad features of the eastern tropical Pacific marine environment follow with an emphasis on Panamá and brief reference to the Galápagos Islands. Numerous marine connections existed across the Central American isthmian region from the Middle (16–15 million years ago [mya]) to Late (7–6 mya) Miocene epochs (Coates and Obando 1996). At that time, present day eastern Pacific and western Atlantic tropical regions shared a common marine biota. Caribbean reef corals occurred in the upper reaches of today's Gulf of California during the early Pliocene, and Indo-Pacific corals were widely distributed throughout the tropical western Atlantic from middle Miocene to early Pliocene times (Budd 1989). By the Middle to Late Pliocene, inter-oceanic connections were reduced to three corridors, and the emerging isthmus became an ecological barrier to the interchange of reef-associated species. The first complete closure of the Pacific-Caribbean marine connection occurred during the interval 3.5–3.1 mya. Subsequently, eastern Pacific and western Atlantic coral reefs evolved in isolation, and precious few species are now shared between these regions. The eastern tropical Pacific today is the most isolated coral reef region in the world (Dana 1975, Grigg and Hey 1992, Veron 1995).

The relatively cool waters of the eastern tropical Pacific are due to the influence of eastern boundary currents, namely the south-southwest flowing California Current, the north-northwest flowing Peru Current, and local upwelling centers (Fiedler 1992, Strub et al. 1998). The northeast and southeast trade winds drive the California and Peru Currents toward the west, and these merge with the North and South Equatorial Currents (NEC, SEC), respectively. Off western Mexico, mean annual sea surface temperatures increase from 18°C to 28°C over 16° of latitude (from 28° to 12°N), and along coastal Peru, Ecuador, and Colombia from 19°C to 27°C over an equal latitudinal tract (12°S to 4°N). The warmest eastern Pacific region, with mean annual surface temperatures of 25°C–28°C, occurs from mid coastal Ecuador to central Mexico, between 3°S and 21°N. The Uva coral reef is located in the thermally stable Gulf of Chiriquí, Panamá, near 8°N (fig. 8.1). This area in western Panamá is located on the leeward side of the Tabasará mountain range, which blocks the northeast trades and thus prevents upwelling.

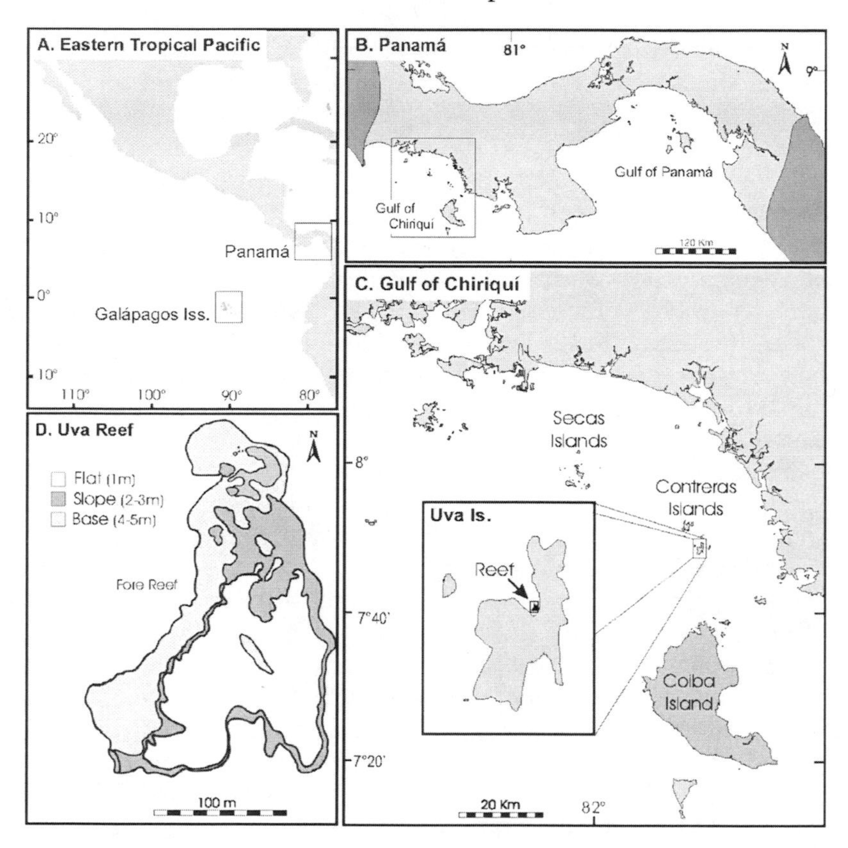

FIGURE 8.1 Locator maps of study sites in Panamá and the Galápagos Islands (A), and the Uva Island coral reef in the Gulf of Chiriquí (B, C). Also shown is a planar view of the Uva reef and its three principal biotic zones (D).

The main coastal upwelling areas in the eastern tropical Pacific are centered in the Gulf of Tehuantepec off southern Mexico, the Gulf of Papagayo astride northwestern Costa Rica and southern Nicaragua, and the Gulf of Panamá lying east of the Gulf of Chiriquí. These areas experience seasonal upwelling during the boreal winter season when the frontal passages drive strong wind jets through the three Middle American corridors of low topographic relief. During upwelling, wind-driven nutricline shoaling stimulates phytoplankton production.

The Intertropical Convergence Zone (ITCZ), the principal climatic influence in Central America, forms where the northeast and southeast trades converge between about 5° and 15°N. This weather system is characterized by light and variable winds and heavy rainfall. The asymmetry in the wind drives a secondary upwelling along the northern ITCZ. This drives an eastward current approximately

8°N, the North Equatorial Countercurrent (NECC), which sets up between the NEC and SEC. Several studies indicate that the NECC is an important vehicle for the dispersal of central Pacific tropical larvae into the eastern Pacific (Emerson and Chaney 1995, Lessios 1998, Reyes Bonilla and López Pérez 1998). When the ITCZ is at its most southerly location off Ecuador, in late-December through April, the dry season prevails in Panamá and the wet season in the Galápagos. In the Gulf of Chiriquí, Panamá the wet season extends from May to November when 90% of the 2,600 mm of annual rain falls. Interannual variability is relatively low, ranging from 1,681 to 3,648 mm over a 20-year period from 1951 to 1970. Mean monthly cloud cover is <10% in the dry season and light and pocilloporid coral skeletal growth rates are highest (Glynn 1977a). In the Galápagos, mean annual rainfall is only about 500 mm, with 70% precipitation occurring from January through May. Annual precipitation is highly variable in the Galápagos, ranging from <100 mm per year to >2,500 mm during warm El Niño events.

Eastern Pacific tropical waters are subject to pronounced interannual variability caused by the 3- to 6-year oscillation of the El Niño-Southern Oscillation (ENSO), a manifestation of the coupled ocean-atmosphere system of the tropical Pacific. Prolonged elevated surface water temperatures (SSTs) associated with ENSO have resulted in two and four coral bleaching and mortality events, respectively, in the Gulf of Chiriquí, Panamá, and the Galápagos Islands over the past 30 years (Podestá and Glynn 2001). Elevated SST anomalies were higher in the Galápagos than in Panamá during this period and coral populations in the former area experienced coral mortality reaching 95%–99% in 1983.

While data on the water quality characteristics of the Gulf of Chiriquí are sparse, due to its remote location, recent sampling on a cross-shelf transect is providing information on wet- and dry-season differences. The seasonal differences at the non-upwelling Uva reef site are less than the upwelling sites, but do result in ecologically significant effects. Sea surface temperature, salinity, and water transparency are highest in the dry season (fig. 8.2, Glynn 1977a, Kwiecinski and Chial 1983, Glynn et al. 2001). Inorganic nutrients and chlorophyll a are higher during the wet season, largely due to increased river discharge. While fish larval abundances were also high in the wet season, zooplankton biomass was highest in the dry season.

Astride the equator, mean annual sea surface temperatures in the Galápagos Islands are the lowest globally at this latitude, ranging from 24°C to 25°C. Three main seasonally-varying current systems converge at the Galápagos Islands: (1) the warm southwest-flowing Panama Current; (2) the cool northwest-flowing Peru Oceanic Current; and (3) a cooler east-flowing subsurface Equatorial Undercurrent, which surfaces along the western and southern

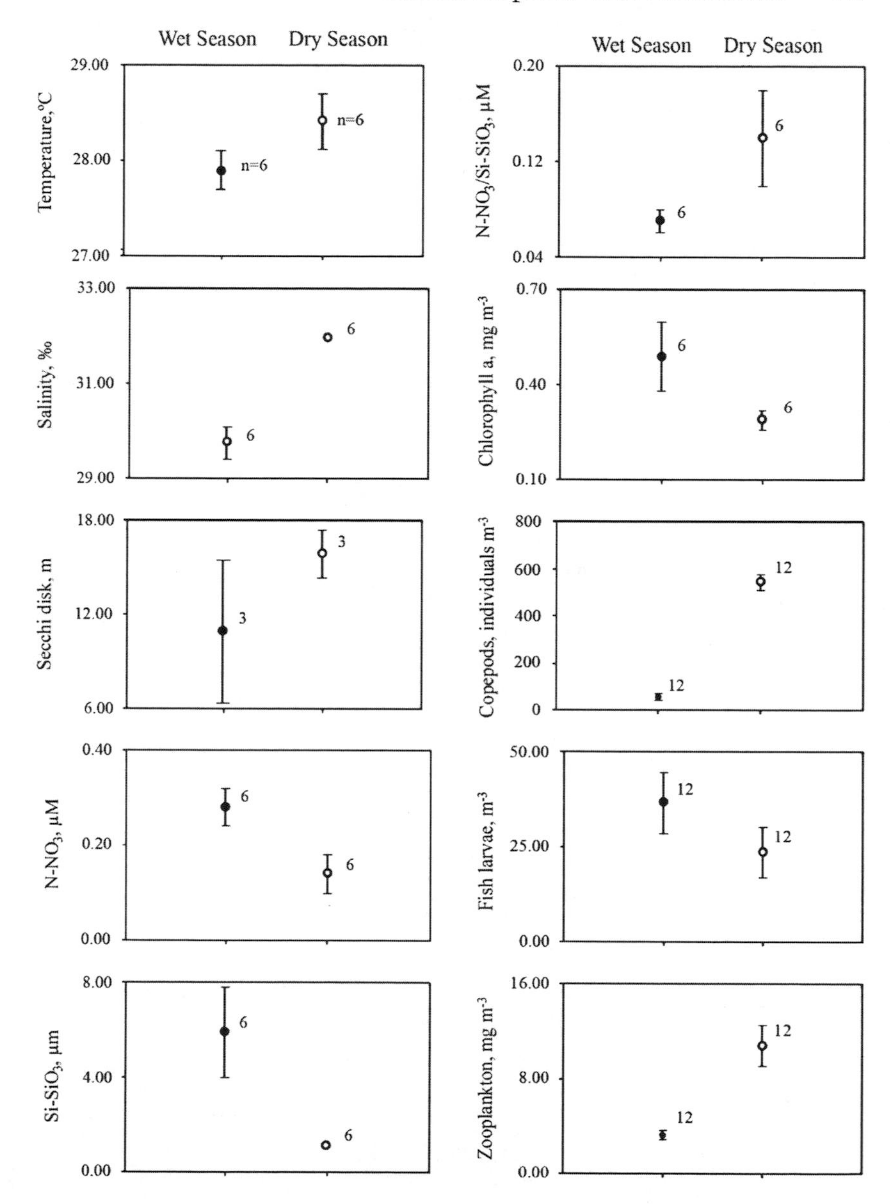

FIGURE 8.2 Mean seasonal surface water quality and biological characteristics along a cross-shelf transect (82.00°W, 7°45′N-8°12′N) in the Gulf of Chiriquí, Panamá. SEMs and number of samples are denoted for each collection. Sampling periods: 16–21 August 1999 (wet season), 22 February–1 March 2000 (dry season) using zooplankton oblique net tows from 20 m to surface. Adapted from Macias Mayorga 2002.

margins of the archipelago (Chavez and Brusca 1991). The Equatorial Undercurrent and elevated concentrations of iron originating from the Galápagos platform (Martin et al. 1994) generate a highly productive upwelling system. Coral reefs occurred mainly on the eastern

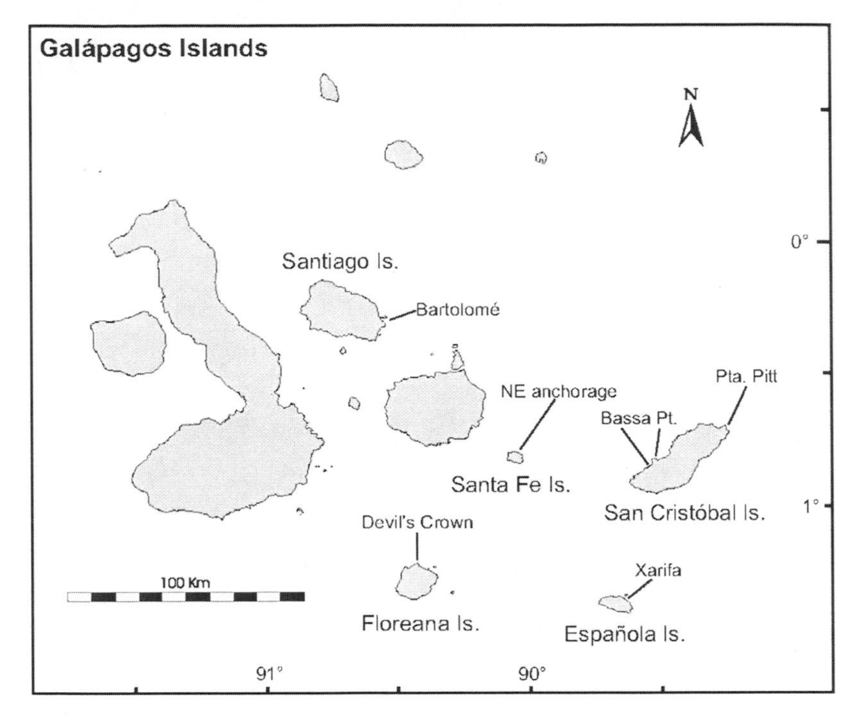

FIGURE 8.3 Galápagos Islands showing locations of six coral-reef study sites. Benthic cover was sampled at two areas on either side of Bassa Point.

sides of the southern and central islands and at the northern islands, under the influence of the Panama Current from January to April and the Peru Oceanic Current during the remainder of the year. Food-web changes following El Niño events were quantified at six coral community sites: Devil's Crown, Bartolomé, NE anchorage at Santa Fe Island, Xarifa, Bassa Point, and Punta Pitt (fig. 8.3).

Biotic Setting

Compared with reef-building coral species richness in the major biogeographic regions, the eastern Pacific zooxanthellate coral fauna is relatively impoverished. Species contour plots denote <50 species in the eastern Pacific, compared with >50 in the western Atlantic and >500 (maximum) in the Indo-West Pacific region (Veron 2000). Recent increases in eastern Pacific coral species, from about 20–25 to 38–42 species, are due to a greater sampling effort and to the naming of new species (Glynn and Ault 2000, Reyes Bonilla 2002). Four species of zooxanthellate hydrocorals (*Millepora*) have also been added to this fauna. However, the number of coral and hydrocoral species present in any given eastern Pacific coral community commonly ranges between only 8 and 12. The eastern Pacific reef fish fauna with 392 species is somewhat lower than in the western Atlantic (444 species), but less diverse compared with South Pacific (519 species) and Great Barrier Reef (1,002 species) faunas (Robertson 1998). Other reef-associated taxa in the eastern Pacific are also relatively depauperate when compared with Western Atlantic and Indo-West Pacific reef biotas (Paulay 1997).

The approximately 2.5 ha Uva reef is one of the best-developed coral reefs in the eastern Pacific. The chief framework builders are branching pocilloporid corals; the coral community consists of 10 zooxanthellate scleractinian and one hydrocoral species. Core drilling has revealed a maximum thickness of 11 m and radiocarbon dating indicates that this reef began growing vertically about 5,600 years ago (Glynn and Macintyre 1977). Significant declines in live coral cover, ranging from 65% to 94%, have occurred during the past 26 years (Wellington and Glynn 2007). These losses have resulted from El Niño warming and associated coral bleaching events and extreme low-tidal exposures. The Uva reef was still in an erosional phase when measured in 2000 (Eakin 2001).

Before the 1982–1983 El Niño disturbance, the five Galápagos coral reefs and one coral community at Bassa Point were among the most diverse and mature in the southern and central archipelago (Glynn and Wellington 1983, Glynn 2003). Pocilloporid corals were the dominant frame-building taxa at all sites except for Bartolomé where massive poritid corals predominated. The number of coral species at each of these sites ranged from four to eight. The Galápagos pocilloporid buildups were youthful when compared with the Uva reef, with vertical thicknesses of 0.5 to 1.0 m and a few hundred instead of thousands of years in age (Macintyre et al. 1992). The horizontal extent of these reefs was also modest, ranging between 1 and 2 ha. All coral reef frameworks described in the 1970s disappeared by the early 1990s, due to El Niño-caused mortality and subsequent bioerosion (Glynn 1994, 2003, Wellington and Glynn 2007).

The Food Web

Methods of Study

The Uva ecosystem-wide food web and corallivore subweb were constructed from long-term day, night, and crepuscular feeding observations on the reef initiated in 1970. Consumers comprising the cryptic subweb were identified from the species that recruited to simulated framework units randomly spaced along the fore reef. These spherically shaped units were composed of pocilloporid coral rubble, each 23–30 liters in volume that were bound in netting and tethered to the bottom. Ten rubble units were collected and sampled after 6, 12,

18, and 24 months from 2002 to 2004. The feeding habits of cryptic consumers were determined from gut content analyses and inferred from the morphology of feeding structures. Traps baited with dead fish flesh also were inserted in the rubble units and natural reef frame to determine the species involved in scavenging.

Respectful of the strict regulations designed to protect coral in the Galápagos Marine Resources Reserve, sampling of coral structures was nondestructive. Four to seven, 10-meter long chain-link transects were established at each of six study sites to determine epibenthic surface cover. These transects covered all coral zones from 0.5 to 8.0 m depth. The same transects were sampled in 1975 and 1976, and re-sampled at least six times from 1984 to 2004 (see Glynn and Wellington 1983, Glynn 1994 for further details). Feeding habits were determined from observations of the visible epibenthos and gut analyses of selected vagile species.

Low Diversity but High Complexity

Despite only two coral species (*Pocillopora damicornis* and *Pocillopora elegans*) contributing importantly to the structure of the Uva reef, all of the trophic guilds found in other high diversity coral ecosystems are also present (fig. 8.4). Moreover, each consumer guild is composed of numerous invertebrate and fish species. Most consumers at trophic levels two and three are residents, closely associated with the reef coral community. Examples of transient consumers are herbivorous sea turtles, carangid invertivores, and a variety of schooling fishes, sharks, groupers, and wahoo, members of trophic level four. While only plants and detritus from trophic level one are shown contributing to the decomposer guild, upon death all organisms at all trophic levels are eventually decomposed. It is recognized that species assignments to particular guilds are not strict. Among herbivores, some species of damselfishes, angelfishes, parrotfishes, and turtles have been observed feeding on metazoans at different times or life history stages. For example, both adult and juvenile parrotfishes (*Scarus ghobban* and *Scarus rubroviolaceus*) occasionally prey on coral tissues and gelatinous zooplankton, including cnidarians, ctenophores, and salps. Since this consumption of animal tissues is sporadic and minimal compared with plants, an omnivore guild is not designated in the generalized food web.

Prominence of corallivores A minimum of twenty invertebrate and fish species consume hard coral tissues on the Uva reef (fig. 8.5). The largest number of corallivores consumes pocilloporid corals. Invertebrate gastropods feeding on *Pocillopora* species are *Quoyula madreporarum* and *Jenneria pustulata*, the latter causing extensive damage and whole-colony mortalities at high population densities

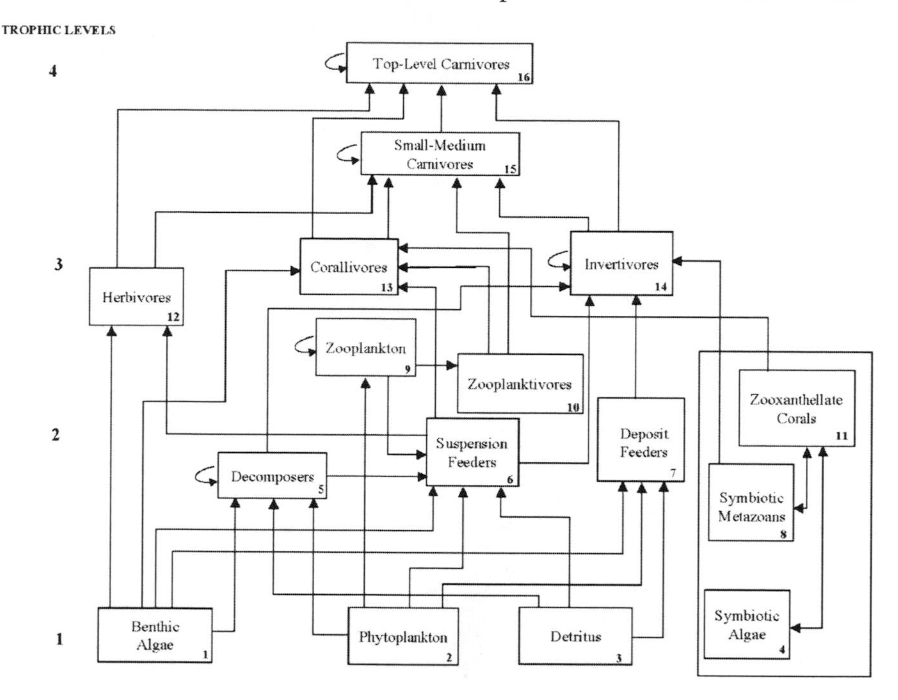

FIGURE 8.4 Conceptual food web of the Uva Island (Panamá) coral reef. Trophic level assignments follow Opitz (1996). Representative guild membership: 1. benthic algae—algal turf, endolithic algae, crustose coralline algae, macroalgae; 2. phytoplankton—femtoplankton to mesoplankton; 3. detritus—particulate organic matter; 4. symbiotic algae—*Symbiodinium* clades A, C, and D; 5. decomposers—microorganisms, meiofauna and microfauna; 6. suspension feeders—sponges, polychaete worms, lithophage bivalves, barnacles; 7. deposit feeders—echiurans, sea cucumbers, shrimp, spiny lobster; 8. symbiotic metazoans—crabs, shrimp; 9. zooplankton—cnidarians, crustaceans, various larval stages; 10. zooplanktivores—cnidarians, damselfishes, manta rays, juvenile parrotfishes; 11. zooxanthellate corals—scleractinian corals and hydrocorals; 12. herbivores—gastropods, echinoids, damselfishes, angelfishes, parrotfishes, kyphosids, acanthurids, turtles; 13. corallivores—hermit crabs, gastropod mollusks, sea stars, pufferfish, parrotfishes; 14. invertivores—polychaete worms, octopuses, hawkfish, jacks (*Caranx speciosus*), triggerfish, wrasses, serranids; 15. small to medium carnivores—polychaete worms, crustaceans (*Hymenocera*, spiny lobster), puffers, balistids, eels, snappers, wrasses, groupers, jacks, sharks; 16. top-level carnivores—sharks, barracuda, wahoo, snappers, groupers, mackerels. Slightly modified after Glynn 2004.

(Glynn 1985). *Acanthaster planci* also preys heavily on *Pocillopora*, but consumes mainly fragmented branches due to protection by symbiotic crustaceans. Fishes from five families prey on *Pocillopora*, with the guineafowl puffer *Arothron meleagris* being the most important (fig. 8.6). However, these puffers bite off only peripheral branches and thus do not kill coral colonies. *Jenneria*, *Acanthaster*, and *Arothron meleagris* cause the highest levels of *Pocillopora* tissue consumption, denoted by thick lines in figure 8.5. Since *Acanthaster*

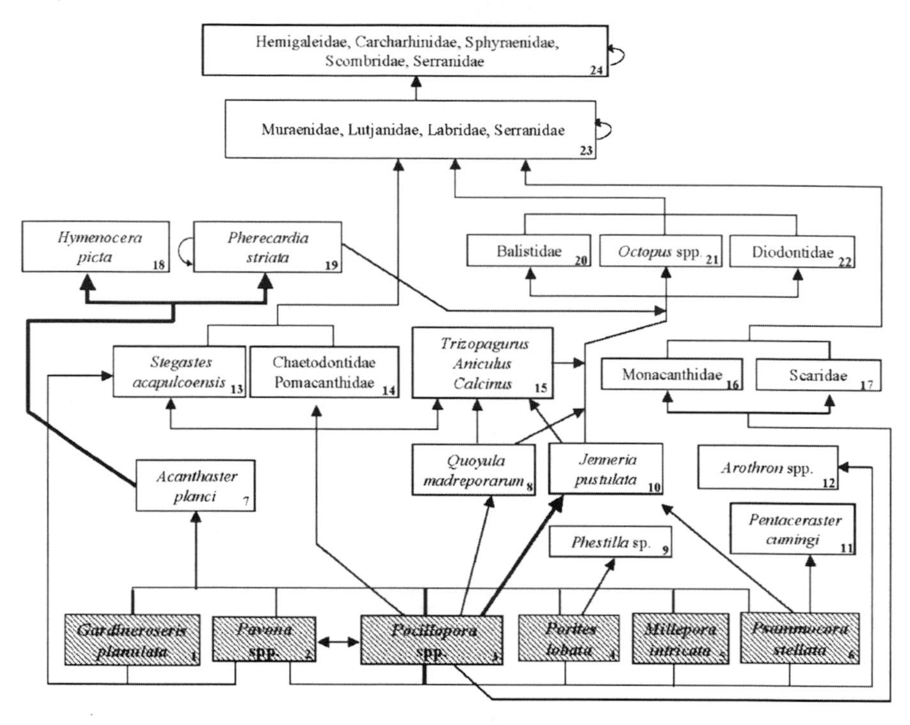

FIGURE 8.5 Uva reef corallivore subweb, illustrating the known coral consumers and their predators. Thick arrows indicate strong pathways, denoting frequent prey consumption. Species identities: 1. *Gardineroseris planulata*; 2. *Pavona* spp. (*Pavona clavus, Pavona gigantea, Pavona varians, Pavona chiriquiensis*); 3. *Pocillopora* spp. (*Pocillopora damicornis, Pocillopora elegans, Pocillopora eydouxi*); 4. *Porites lobata*; 5. *Millepora intricata*; 6. *Psammocora stellata*; 7. *Acanthaster planci*; 8. *Quoyula madreporarum*; 9. *Phestilla* sp.; 10. *Jenneria pustulata*; 11. *Pentaceraster cumingi*; 12. *Arothron* spp. (*Arothron meleagris, Arothron hispidus*); 13. *Stegastes acapulcoensis*; 14. Chaetodontidae and Pomacanthidae (*Johnrandallia nigrirostris, Chaetodon humeralis, Holacanthus passer, Pomacanthus zonipectus*); 15. hermit crabs (*Trizopagurus magnificus, Aniculus elegans, Calcinus obscurus*); 16. Monacanthidae (*Aluterus scriptus, Cantherhinus dumerilii*); 17. Scaridae (*Scarus perrico, Scarus rubroviolaceus*); 18. *Hymenocera picta*; 19. *Pherecardia striata*; 20. Balistidae (*Sufflamen verres, Pseudobalistes naufragium*); 21. *Octopus* spp.; 22. Diodontidae (*Diodon holocanthus, Diodon hystrix*); 23. Muraenidae (*Gymnothorax castaneus, Gymnothorax dovii*), Lutjanidae (*Lutjanus viridis, Lutjanus argentiventris, Lutjanus novemfasciatus*), Labridae (*Bodianus diplotaenia, Novaculichthys taeniourus, Thalassoma grammaticum*), and Serranidae (*Epinephelus panamensis, Epinephelus labriformis*); 24. Hemigaleidae (*Triaenodon obesus*), Carcharhinidae (*Carcharhinus leucas*), Sphyraenidae (*Sphyraena ensis*), Scombridae (*Acanthocybium solandri*), and Serranidae (*Mycteroperca xenarcha*). Slightly modified after Glynn 2004.

also shows a feeding preference for *Gardineroseris planulata*, a strong interaction is indicated. This particular predator–prey interaction was especially pronounced in the 1970s when 30 to 40 individuals of *Acanthaster* were present on the Uva reef (Glynn 1990, Fong

FIGURE 8.6 A Guineafowl pufferfish, *Arothron meleagris*, feeding on the branchtips of *Pocillopora damicornis*, Uva Island coral reef, Panamá, 2 meters' depth, 12 March 2004. Photograph courtesy of M. Wartian.

and Glynn 1998). Finally, the extrusion of mesenterial filaments by *Pocillopora* species onto *Pavona clavus* and *Pavona gigantea* results in the digestion and death of tissues, which may serve as a source of nutrition. The development of sweeper tentacles by *Pavona* reverses this interaction after 1 to 2 months (Wellington 1980).

Predator–prey studies in Panamá have demonstrated that certain reef invertebrates and fishes at higher trophic levels are probably critical in limiting the abundances of adult *Acanthaster* and *Jenneria* (Glynn 1977b, 1984; Oramas 1979). *Acanthaster* abundances appear to be controlled by a harlequin shrimp (*Hymenocera picta*) and a polychaete worm (*Pherecardia striata*). Hermit crabs, octopuses, and fishes (balistids and diodontids) consume *Jenneria*. It is likely that hermit crabs also feed on the egg masses of *Jenneria*, which are deposited on the dead basal branches of pocilloporid corals. No predators of adult *Arothron meleagris* have been observed.

Cryptic consumers The cryptofauna or coelobites, those largely unseen animals living in bioeroded holes, crevices, and other sheltered spaces on coral reefs, may generally contribute proportionately more to the diversity and biomass of reef communities than the visible surface fauna (Ginsburg 1983, Reaka-Kudla 1997). Many of these cryptic species are micropredators that form part of the zooplanktivore, invertivore, and small-to-medium sized carnivore guilds (fig. 8.7). Among such invertebrate predators inhabiting the Uva Island reef are polyclad flatworms, nemertean and polychaete worms, cone and ovulid gastropods, octopuses, cirolanid, and tanaid isopods and stomatopods. Cryptic fish predators include cuskeels, moray eels, snake eels, frogfishes, scorpionfishes, juvenile snappers and groupers, cardinal fishes, squirrelfishes, weed blennies, and soapfishes.

The diverse community of cryptic species occupying the internal surfaces and spaces in coral frameworks suggests a high level of predation both within and immediately surrounding these structures. Major consumer guilds identified include suspension feeders,

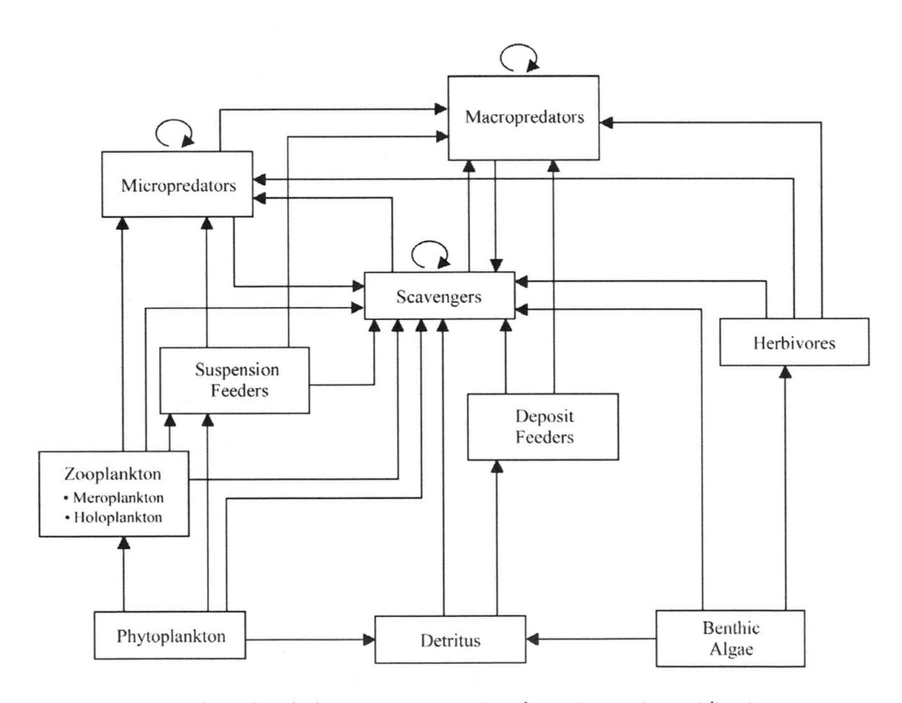

FIGURE 8.7 Trophic subweb depicting community of cryptic species residing in pocilloporid coral frameworks, Uva Island reef, Panamá. Algae: microfilamentous, macroalgae, crustose coralline algae; Herbivores: amphipods, opisthobranchs, limpits, chitons; Deposit feeders: holothurians, polychaetes, sipunculids, echiurans, *Hipponix grayanus*; Suspension feeders: sponges, hydroids, bryozoans, bivalves, vermetids, ophiuroids, serpulid polychaetes, porcellanid crabs, barnacles; Scavengers: *Pherecardia* striata, gastropods, cirolanid isopods, ostracods, hermit crabs, holothurians; Micropredators: polyclad flatworms, isopods, nemerteans, polychaetes, pygmy octopuses; Macropredators: cones, ovulid gastropods, stomatopods, moray eels (Muraenidae), snake eels (Ophichthidae), brotulas (Ophidiidae), pearlfishes (Carapidae), cuskeels (Bythitidae), frogfishes (Antennariidae), squirrelfishes (Holocentridae), scorpionfishes (Scorpaenidae), juvenile groupers (Serranidae), soapfishes (Grammistidae), cardinalfishes (Apogonidae), juvenile snappers (Lutjanidae), blennies (Labrisomidae).

deposit feeders, herbivores, and micro- and macropredators. Various invertebrate taxa comprise the nonpredatory consumer guilds and also dominate the micropredators. While a few invertebrates such as mollusks and crustaceans are macropredators, all of the known cryptic fishes, members of 13 families, are carnivores.

Linkages between surface consumers and the cryptic fauna are numerous and bidirectional. For example, sea turtles (*Lepidochelys divacea* and *Caretta caretta)*, the whitetip reef shark (*Triaenodon obesus*) and balistids (*Sufflamen verres* and *Pseudobalistes naufragium*) break apart pocilloporid frameworks while pursuing sponges, fishes, and various higher invertebrates, respectively (Jiménez 1996–1997, Glynn 2004). Several small opportunistic fishes such as *Halichoeres dispilus, Thalassoma lucasanum, Bodianus diplotaenia,* and *Cirrhitichthys oxycephalus* also are attracted to the afore-noted excavating macropredators and prey on the many species thus exposed. Cryptic herbivores such as *Dolabella auricularium* and *Diadema mexicanum*, scavengers (*Pherecardia striata,* cirolanid isopods, ostracods and gastropods), deposit feeders (*Euapta godeffroyi* and *Holothuria impatiens*), suspension feeders (bivalves and porcellanid crabs), and micro- and macropredators (pygmy octopuses, eels, stomatopods, and squirrel fishes) emerge from coral frameworks, often at night, to prey on organisms in the water column and those inhabiting reef surfaces.

Spatial and Temporal Variation

Spatial Variation

Some potentially influential eastern Pacific coral reef associates are absent or present at low abundances depending upon location. For example, *Acanthaster planci* is usually absent from coral communities and reefs present in upwelling centers or subject to seasonally cool water conditions such as the Gulf of Panamá and the central and southern Galápagos Islands. In Panamá, coral species present in habitats with *Acanthaster* demonstrated significantly smaller colony sizes and higher proportions of dead colonies than the same species in similar habitats without *Acanthaster* (Glynn 1987). The gastropod corallivore *Jenneria pustulata* is rare in the Galápagos Islands, but can attain high population densities in the Gulf of Chiriquí. Additionally, the bioeroding echinoid *Eucidaris galapagensis* inhabits only some offshore islands (Galápagos Islands and Cocos Island). *Eucidaris* is an omnivore, consuming algae and coral tissues.

Spatial differences in abundances are also evident at small scales within given reefs, resulting in varying patterns of predation and effects on particular prey populations. On the Uva reef, *Acanthaster* forages on the fore reef and reef base, largely sparing corals on the reef flat and in back-reef zones. Further, massive coral species surrounded by branching pocilloporid corals are protected from *Acanthaster* attack by the crustacean guards and nematocysts present in the branching coral barrier. Large, transient fish carnivores also concentrate their feeding in fore-reef and reef-base zones (Glynn 2004).

The feeding preferences, territorial behavior, and depth distribution of damselfishes can affect the distribution and relative abundances of coral species. For example, Wellington (1982) found that the Acapulco Gregory *Stegastes acapulcoensis* feeds sparsely on branching pocilloporid corals, consuming approximately 2% of their surface cover, but feeds heavily on the smooth surfaces of pavonid corals. In the Pearl Islands (Gulf of Panamá), *Stegastes* is abundant in shallow reef zones where it shelters in pocilloporid corals. In this habitat the damselfish reduces the entry of fish herbivores and corallivores, which allows the proliferation of pocilloporid corals. Also, the damselfish in shallow *Pocillopora* zones kills *Pavona* species, up to 59% of polyp surfaces. In deep-reef zones where shelter refuges are rare and damselfish abundances low, pocilloporid corals receive less protection and pavonid corals are generally the predominant genera.

Temporal Variation

Uva reef Several reef species have demonstrated large variations in abundance since the early 1970s, but only some of these changes have been documented. At the Uva reef, large numbers of *Diadema mexicanum*, with densities as high as 60–80 ind m^{-2} in some reef zones, were present from 1985 to 1996 (Glynn 1988b, Eakin 2001). Filamentous and crustose coralline algae that colonized large tracts of pocilloporid corals killed during the 1982–1983 ENSO were then grazed by *Diadema*, resulting in extensive erosion and loss of reef frameworks. For reasons unknown, *Diadema* abundances have declined markedly in the late 1990s to pre-1983 levels (0–1 ind m^{-2}).

Two corallivores that have had varying effects on Uva coral populations are the sea star *Acanthaster planci* and the ovulid gastropod *Jenneria pustulata*. When abundant, *Acanthaster* densities reached 15–20 ind ha^{-1}. Unlike many Indo-Pacific reef areas, *Acanthaster* outbreaks have not been observed in the eastern Pacific. *Jenneria* mean population densities during 1982 were 15–28 ind m^{-2}. *Gardineroseris planulata* were preyed on heavily by *Acanthaster*, as were broken branches of pocilloporid corals. *Jenneria* preyed almost exclusively on pocilloporid colonies, with breeding aggregations consuming whole large colonies. Both species were at low abundances in 2004, with approximately 1 *Acanthaster* ha^{-1} and 0–1 *Jenneria* m^{-2}, and thus have little effect on Uva reef corals. Resident reef fish abundances, including herbivores, corallivores, and carnivores, have been relatively stable during the past 34 years (P. Glynn, unpublished data), but transient manta rays and sharks have undergone apparent declines (Glynn 2004).

The Galápagos Islands Notable changes have occurred in the structure of coral community food webs in the Galápagos Islands during the past quarter century. All coral community study sites monitored since the mid 1970s suffered nearly total coral loss during and following the 1982–1983 El Niño event (Glynn 1994). The predisturbance dimensions of the coral framework structures at these sites suggest young coral community ages of a few to several centuries (Glynn and Wellington 1983, Macintyre et al. 1992). Unlike Panamanian coral communities, which as of 2003–2004 demonstrated significant recovery by means of asexual tissue regeneration and sexual recruitment, all but a few Galápagos coral communities are still highly degraded with minimal live coral cover. As of 2002, 15 of 17 study sites still have very low coral cover, demonstrating no significant recovery since 1983 (Glynn 2003). Indeed, they are now probably best designated as hard bottom algal turf and sea urchin barrens.

A pre- and post-El Niño analysis of the trophic structure of the Devil's Crown coral reef showed that the mean coral cover of 37.1% (SD = 20.8, n = 7 transects) in 1975 decreased significantly to near-zero values in 1989 and 2002 (fig. 8.8). Crab and shrimp obligate mutualists that inhabit *Pocillopora* colonies also disappeared with their hosts. A second significant change was an increase in herbivore populations, which ranged from 0% of the epibenthic cover in 1975 to 6.0% (SD = 3.0, n = 7) in 1989. This increase was due primarily to the echinoid *Lytechinus semituberculatus* with a mean population density of 5 ind m^{-2}. Although the omnivorous *Eucidaris* numbers declined steadily from 10–25 ind m^{-2} immediately following the 1982–1983 El Niño, the destructive feeding behavior of fewer sea urchins (2–4 ind m^{-2}) in subsequent years was still sufficient to sustain continued erosion of the reef frame.

The Bartolomé coral reef, constructed chiefly of the massive coral *Porites lobata*, also demonstrated a significant decline in live coral cover following the 1982–1983 El Niño event (fig. 8.9). The only other significant change in the abundance of a consumer guild was an increase in *Eucidaris* in 1988. Significant declines in zooxanthellate coral cover also occurred after the 1982–1983 El Niño event in all coral communities at the Xarifa, Punta Pitt, Bassa Point, and Santa Fe study sites. Benthic algal cover at Punta Pitt in 2003 was high at 86%, significantly higher than in 1976 or 1989 when it was 55% and 76%, respectively.

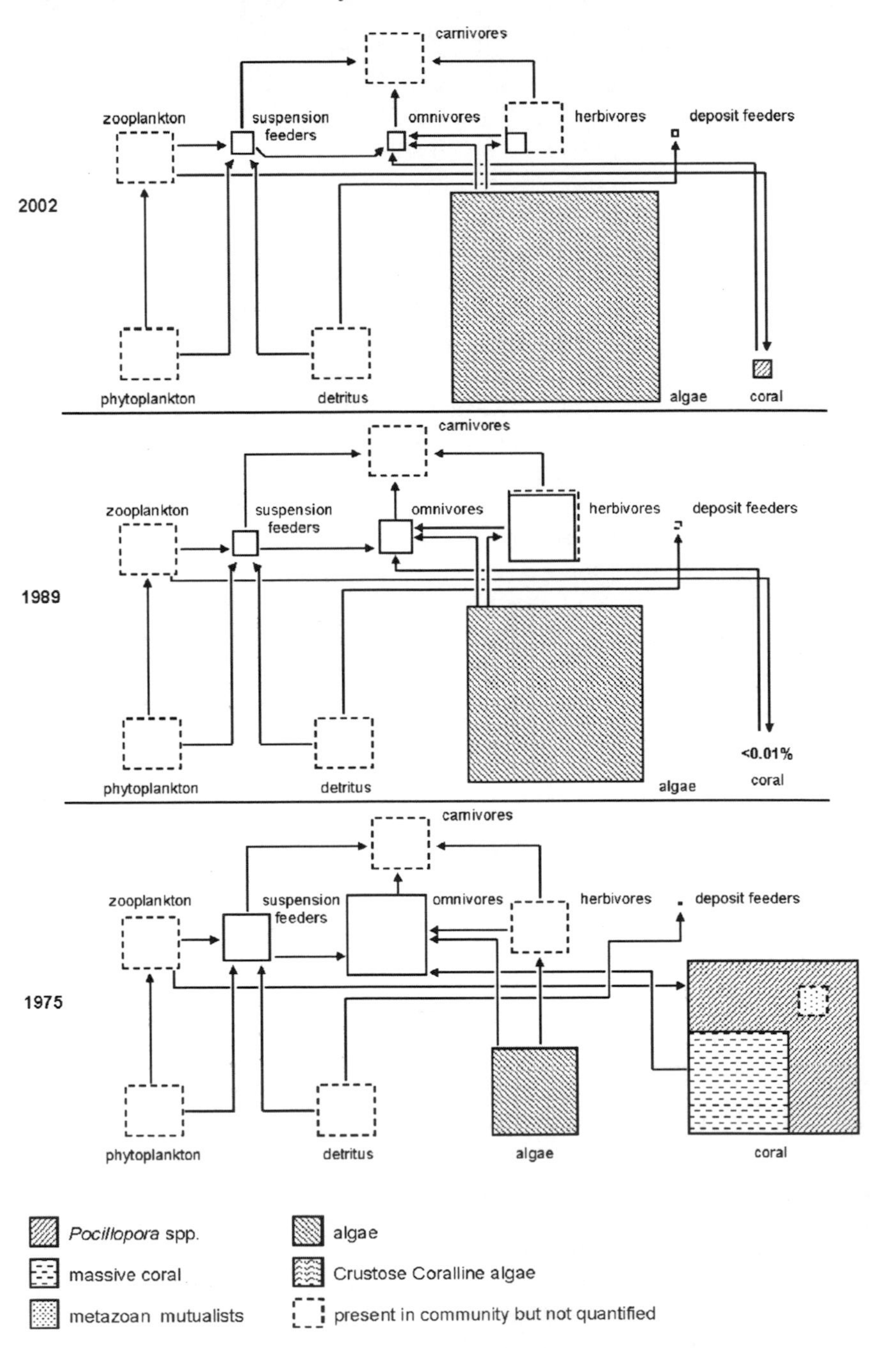

FIGURE 8.8 Major food web relationships at the Devil's Crown (Floreana Island) coral community over a 27-year period. In the 1970s this community was dominated by branching coral species, principally *Pocillopora elegans*, which in 1975 contributed 22.4% (±24.6 SD) to the mean epibenthic cover. Filled box areas are proportional to the surface cover of the sampled benthos. Broken box compartments denote the presence of other trophic categories that were not quantitatively sampled.

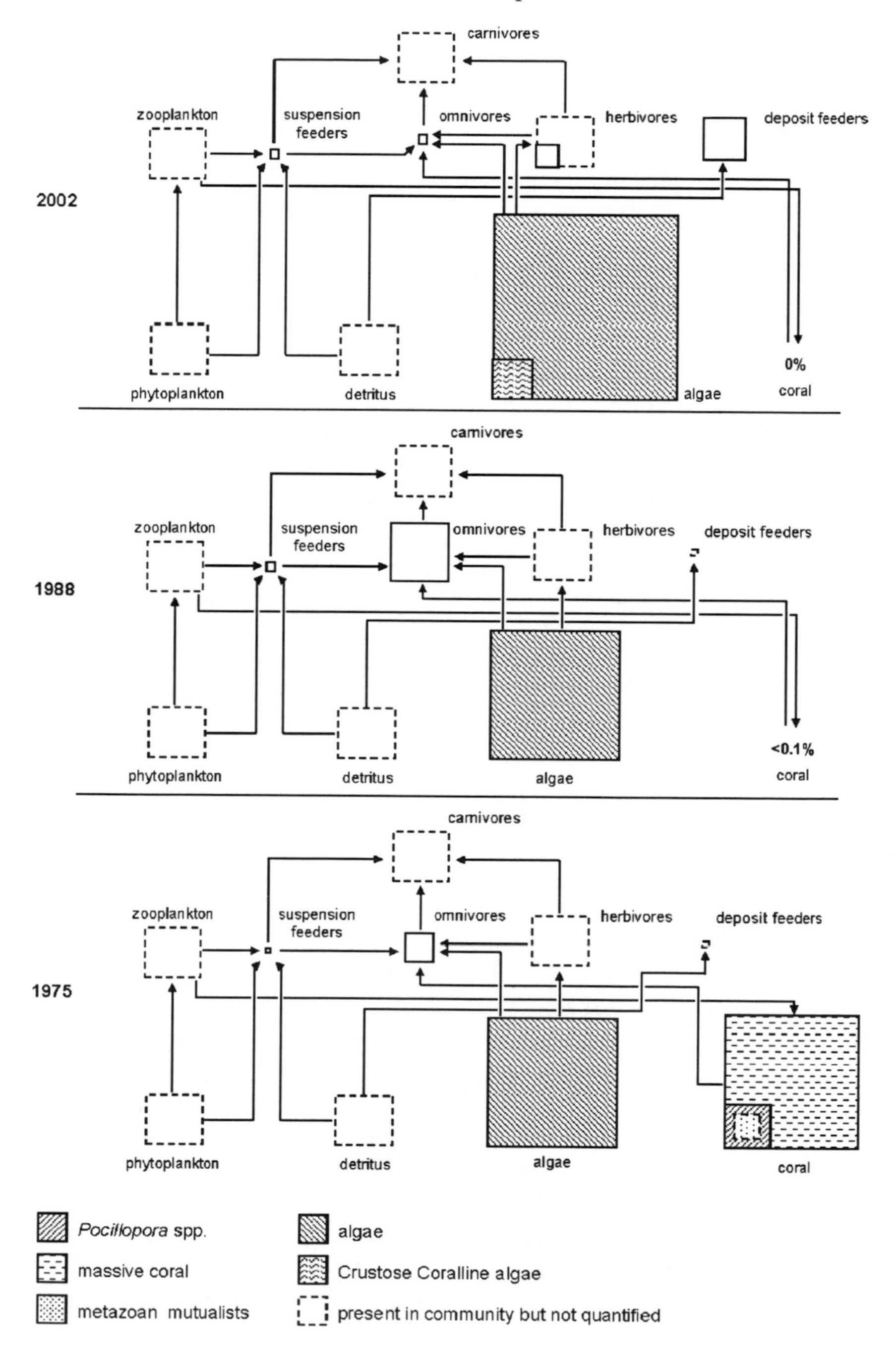

FIGURE 8.9 Major food-web relationships at the Bartolomé (Santiago Island) coral community over a 28-year period. In the 1970s this community was dominated by massive coral species, principally *Porites lobata*, which in 1975 contributed 20.1% (±7.4 SD) to the mean epibenthic cover. Filled box areas are proportional to the surface cover of the sampled benthos. Trophic box categories as in figure 8.7.

Human Influences

Uva Reef

While it is clear that the abundances of several reef-associated species show marked inter-annual fluctuations, it is not always possible to determine if these variations are due to environmental or direct anthropogenic influences. In the majority of cases, it is possible to note only correlations with presumed causative agents. Marketable species that are now uncommon-to-rare on Panamanian reefs include conchs, pearl oysters, spiny lobsters, sharks, and turtles. It is easier to attribute low numbers of the above to exploitation, but some species that have declined recently, such as Manta rays (*Manta birostris*) with wingspans of 3–4 m, were commonly present at the Uva reef where in the 1970s they fed in the upper water column along the fore reef (Glynn 2004). Their decline in Panamá during the past two decades does not seem to be related to over-fishing since commercial or artisanal fishers do not target them, and they are rarely captured as by-catch. There is a large fisheries for mantas and mobulas off the Pacific coast of Mexico, but no definitive fishery information is available (T.B. Clark, personal communication). Owing to their low fecundity (1 pup every 1–3 years), slow growth (longevity 50–100 years), and limited migration, *M. birostris* would appear to be highly susceptible to overfishing.

Based on visual fish surveys performed across three reef zones (P. Glynn, unpublished data), the abundances of the noncryptic, diurnally-active reef fishes (TL $\geq$ 15 cm) have changed little since they were first quantified in the early 1970s. In 2000–2002, as in the early 1970s, omnivorous balistids were among the most abundant taxa observed, with mean densities of 105 ind ha^{-1}. These were followed in abundance by herbivorous scarids at 100 ind ha^{-1}, the corallivorous guinea fowel puffer (*Arothron meleagris*) at 65 ind ha^{-1}, carnivorous snappers at 40 ind ha^{-1}, and herbivorous acanthurids at 34 ind ha^{-1}. I speculate that low fishing pressure has contributed to this relative constancy in reef fish abundances over the last 30+ years. Certainly, most fishing effort in the Gulf of Chiriquí is conducted on relatively deep (20–100 m) offshore banks and rock reefs where snapper and grouper species are the primary targets.

The Uva Island coral reef is located within the bounds of the Coiba National Park, a wildlife protected area established by the Panamanian government in 1991. This area is relatively pristine, due largely to the presence of a penal colony and custodial police force on Coiba Island. This has resulted in limited access for more than 80 years. With the closure of the penal colony and insufficient funds to enforce protection, it is likely that illegal logging, over-fishing, and other destructive activities will lead to major disturbances to coral reefs in this newly created park (Maté 2003).

The Galápagos Islands

Under the stewardship of the Galápagos National Park Service, with scientific support from the Charles Darwin Research Station, presently most coral community sites are adequately protected and managed. Among the six study sites only Devil's Crown and Bartolomé are located within Marine National Park Zones, which prohibit the removal of natural resources. Nonetheless, I have not observed any significant human-related physical or biotic disturbances at these sites during the past 30 years. While the Galápagos Marine Resources Reserve is internationally recognized as a model in conservation efforts, the high immigration rate of humans and their effects on the environment are cause for concern (chapter 6).

As in Panamá, fishes associated with coral communities have not exhibited extreme variation in abundance or local extinctions in recent years. Fish species affected by the 1982–1983 and 1997–1998 El Niño events in the Galápagos have generally recovered to moderate population densities (Jennings et al. 1994). Overfishing of lobsters during the past several decades has greatly reduced the abundances of these scavengers in Galápagos coral communities. Recently, the sea cucumber *Stichopus fuscus* has come under heavy exploitation, with an estimated island-wide removal in 1995 of 8–12 million individuals by 850–900 divers (Glynn 2003). Since the bulk of these deposit feeders is being removed from non-coral habitats, this large-scale extraction may not result in direct effects to coral communities.

In the eastern Pacific, notably in the Galápagos Islands, low coral recruitment and intense bioerosion have caused widespread loss of reef frameworks. Considering the importance of coral frameworks in providing habitat space for surface- and cryptic-dwelling reef species, it is apparent that coral reef restoration efforts should utilize methods that will promote the re-constitution of reef structures. The re-seeding of coral communities with fragments from branching coral colonies has been shown to initiate framework building. For example, fragments from ramose *Acropora* and *Pocillopora* colonies have been used successfully in reef restoration projects in the Caribbean (Bowden-Kerby 2001) and eastern Pacific (Guzmán 1991), respectively. To initiate framework development of degraded eastern Pacific coral reefs, pocilloporid fragments offer several advantages: (a) availability, pocilloporid reefs predominate in this region; (b) high survivorship on firm substrates; (c) high growth rate; (d) the likelihood of continued recruitment through asexual fragmentation; (e) high topographic complexity; and (f) the production of trophic resources, such as coral tissues, lipid bodies, and mucus that are utilized by numerous associated species, some of which are obligate symbionts.

Conclusions

Two conditions seem to contribute to the highly complex nature of low coral diversity Uva reef food webs. These relate to the porous nature of pocilloporid reef frameworks, which are constructed of interlocking vertically oriented branches, and the large numbers of species that inhabit and feed on pocilloporid corals. The quasi-open structure of pocilloporid frameworks permits vigorous circulation within the upper reef strata and for varying flow rates and sediment deposition at the reef base. This creates microhabitat differences, which offer a range of ecological niches for diverse reef taxa.

Several recent workers have presented evidence that phase shifts in modern coral reefs and other coastal food webs have been caused by overfishing during the past several centuries (Jackson et al. 2001, Pandolfi et al. 2003). These studies hypothesized that the removal of grazing fishes and invertebrates by overexploitation has resulted in the proliferation of macroalgae, sponges, and other epibenthos that can out-compete corals for suitable substrate space. Prolonged elevated sea temperatures associated with recent El Niño events caused the ecological changes and coral reef collapse on Panamanian and Galápagos coral reefs (Glynn 2003). Coral reefs in Panamá and the Galápagos Islands were vibrant, actively growing structures before the 1982–1983 El Niño event.

A challenge facing coral reef trophic studies centers on elucidating the role of the cryptic reef fauna in coral reef growth and recovery. An effort should be directed toward documenting the food-web structure and feeding rates of cryptic reef consumers. An important aspect of this work will be to determine the effect of cryptic consumers on organisms recruiting to coral communities. Preliminary indications suggest that this cryptic community acts as a filter, allowing the recruitment of select species that may strongly influence subsequent community development.

Most changes observed in these reefs relate to recent declines in reef-coral abundances and sea warming events associated with El Niño-Southern Oscillation activity (Hoegh-Guldberg 1999, Wilkinson 2000). Field and laboratory studies have demonstrated that zooxanthellate coral mortality in many areas is due to prolonged exposure to elevated sea temperatures with high irradiance also a factor in some instances (Brown 1997, Fitt et al. 2000). Evidence is mounting that global warming is raising the baseline of tropical ocean temperatures and thus contributing to higher stressful thermal levels during ENSO events.

References

Bowden-Kerby, A. 2001. Low-tech coral reef restoration methods modeled after natural fragmentation processes. Bulletin of Marine Science **69**:915–931.

Brown, B.E. 1997. Coral bleaching: causes and consequences. Coral Reefs **16**:S129–S138.

Budd, A.F. 1989. Biogeography of Neogene Caribbean reef corals and its implications for the ancestry of eastern Pacific reef corals. Association of Australasian Palaeontologists, Memoirs **8**:219–230.

Chavez, F.P., and R.C. Brusca. 1991. The Galápagos Islands and their relation to oceanographic processes in the tropical Pacific. Pages 9–33 *in* M. J. James, editor. Galápagos Marine Invertebrates: taxonomy, biogeography, and evolution in Darwin's Islands. Plenum Press, N.Y.

Coates, A.G., and J.A. Obando. 1996. The geologic evolution of the Central American Isthmus. Pages 9–33 *in* J.B.C. Jackson, A.F. Budd, and A.G. Coates, editors, Evolution and environment in tropical America. University of Chicago Press, Chicago.

Dana, T.F. 1975. Development of contemporary eastern Pacific coral reefs. Marine Biology **33**:355–374.

Eakin, C.M. 2001. A tale of two ENSO events: carbonate budgets and the influence of two warming disturbances and intervening variability, Uva Island, Panama. Bulletin of Marine Science **69**:171–186.

Emerson, W.K., and H.W. Chaney. 1995. A zoogeographic review of the Cypraeidae (Mollusca: Gastropoda) occurring in the eastern Pacific Ocean. The Veliger **38**:8–21.

Fiedler, P.C. 1992. Seasonal climatologies and variability of eastern tropical Pacific surface waters. National Oceanographic and Atmospheric Administration Technical Report, National Marine Fisheries Service **109**:1–65.

Fitt, W.K., F.K. McFarland, M.E. Warner, and G.C. Chilcoat. 2000. Seasonal patterns of tissue biomass and densities of symbiotic dinoflagellates in reef corals and relation to coral bleaching. Limnology and Oceanography **45**:677–685.

Fong, P., and P.W. Glynn. 1998. A dynamic size-structured population model: does disturbance control size structure of a population of the massive coral *Gardineroseris planulata* in the eastern Pacific? Marine Biology **130**:663–674.

Ginsburg, R.N. 1983. Geological and biological roles of cavities in coral reefs. Pages 148–153 *in* D.J. Barnes, editor, Perspectives on coral reefs. Brian Clouston Publisher, Manuka, Australia.

Glynn, P.W. 1977a. Coral growth in upwelling and nonupwelling areas off the Pacific coast of Panama. Journal of Marine Research **35**:567–585.

Glynn, P.W. 1977b. Interactions between *Acanthaster* and *Hymenocera* in the field and laboratory. Proceedings of the Third International Coral Reef Symposium **1**:209–215.

Glynn, P.W. 1984. An amphinomid worm predator of the crown-of-thorns sea star and general predation on asteroids in eastern and western Pacific coral reefs. Bulletin of Marine Science **35**:54–71.

Glynn, P.W. 1985. Corallivore population sizes and feeding effects following El Niño (1982–83) associated coral mortality in Panamá. Proceedings of the Fifth International Coral Reef Congress **4**:183–188.

Glynn, P.W. 1987. Some ecological consequences of coral-crustacean guard mutualisms in the Indian and Pacific Oceans. Symbiosis **4**:301–324.

Glynn, P.W. 1988a. Predation on coral reefs: some key processes, concepts and research directions. Proceedings of the Sixth International Coral Reef Symposium **1**:51–62.

Glynn, P.W. 1988b. El Niño warming, coral mortality and reef framework destruction by echinoid bioerosion in the eastern Pacific. Galaxea **7**:129–160.

Glynn, P.W. 1990. Coral mortality and disturbances to coral reefs in the tropical eastern Pacific. Pages 55–126 *in* P.W. Glynn editor, Global ecological consequences of the 1982–83 El Niño-Southern Oscillation. Elsevier Oceanographic Series **52**, Amsterdam.

Glynn, P.W. 1994. State of coral reefs in the Galápagos Islands: natural vs. anthropogenic impacts. Marine Pollution Bulletin **29**:131–140.

Glynn, P.W. 2003. Coral communities and coral reefs of Ecuador. Pages 449–472 *in* J. Cortés, editor. Latin American coral reefs. Elsevier, Amsterdam.

Glynn, P.W. 2004. High complexity food webs in low-diversity eastern Pacific reef-coral communities. Ecosystems **7**:358–367.

Glynn, P.W., and Ault, J.S. 2000. A biogeographic analysis of the far eastern Pacific coral reef region. Coral Reefs **20**:1–20.

Glynn, P.W., and Macintyre, I.G. 1977. Growth rate and age of coral reefs on the Pacific coast of Panamá. Proceedings of the Third International Coral Reef Symposium **2**:251–259.

Glynn, P.W., J.L. Maté, A.C. Baker, and M.O. Calderón. 2001. Coral bleaching and mortality in Panamá and Ecuador during the 1997–1998 El Niño-Southern Oscillation event: spatial/temporal patterns and comparisons with the 1982–1983 event. Bulletin of Marine Science **69**:79–109.

Glynn, P.W., and G.M. Wellington. 1983. Corals and coral reefs of the Galápagos Islands. University of California Press, Berkeley.

Grigg, R.W., and R. Hey. 1992. Paleoceanography of the tropical eastern Pacific. Science **255**:172–178.

Guzmán, H.M. 1991. Restoration of coral reefs in Pacific Costa Rica. Conservation Biology **5**:189–195.

Hoegh-Guldberg, O. 1999. Coral bleaching, climate change and the future of the world's coral reefs. Marine and Freshwater Research **50**:839–866.

Jackson, J.B.C., M.X. Kirby, W.H. Berger, K.A. Bjorndal, L.W. Botsford, B.J. Bourque, C.B. Lange, H.S. Lenihan, J.S. Pandolfi, C.H. Peterson, R.S. Steneck, M.J. Tegner, and R.R. Warner. 2001. Historical overfishing and the recent collapse of coastal ecosystems. Science **293**:629–638.

Jennings, S., A.S. Brierley, and J.W. Walker. 1994. The inshore fish assemblages of the Galápagos Archipelago. Biological Conservation **70**:49–57.

Jiménez, C. 1996–1997. Coral colony fragmentation by whitetip reef sharks at Coiba Island National Park, Panamá. Revista de Biología Tropical **44/45**:698–700.

Karlson, R.H. 1999. Dynamics of coral communities. Kluwer Academic Publishers, Dordrecht, Germany.

Kwiecinski, B., and B. Chial Z. 1983. Algunos aspectos de la oceanografía del Golfo de Chiriquí, su comparación con el Golfo de Panamá. Revista de Biología Tropical **31**:323–325.

Lessios, H.A., B.D. Kessing, and D.R. Robertson. 1998. Massive gene flow across the world's most potent marine biogeographic barrier. Proceedings of the Royal Society of London B **265**:583–588.

Macias Mayorga, D.M. 2002. Estructura trófica del plancton en el Pacífico de Panamá. Thesis, Escuela de Biología, Facultad de Ciencias Naturales, Exactas y Tecnología, Universidad de Panamá, Panamá.

Macintyre, I.G., P.W. Glynn, and J. Cortés. 1992. Holocene reef history in the eastern Pacific: mainland Costa Rica, Caño Island, Cocos Island, and Galápagos Islands. Proceedings of the Seventh International Coral Reef Symposium **2**:1174–1184.

Martin, J.H., K.H. Coale, and 42 coauthors. 1994. Testing the iron hypothesis in ecosystems of the equatorial Pacific Ocean. Nature **371**:123–129.

Maté, J.L. 2003. Corals and coral reefs of the Pacific coast of Panamá. Pages 387–417 *in* J. Cortés, editor. Latin American Coral Reefs, Elsevier, Amsterdam.

McClanahan, T. 2004. Food webs of shallow subtidal marine ecosystems. Ecosystems **7**:321–322.

Opitz, S. 1996. Trophic interactions in Caribbean coral reefs. ICLARM Technical Report 43.

Oramas N.F.A. 1979. Estudio ecológico del *Jenneria pustulata* (Mollusca, Gastropoda). Thesis, Escuela de Biología, Universidad de Panamá, Panamá.

Pandolfi, J.M., R.H. Bradbury, E. Sala, T.P. Hughes, K.A. Bjorndal, R.G. Cooke, D. McArdle, L. McClenachan, M.J.H. Newman, G. Paredes, R.R. Warner, and J.B.C. Jackson. 2003. Global trajectories of the long-term decline of coral reef ecosystems. Science **301**:955–958.

Paulay, G. 1997. Diversity and distribution of reef organisms. Pages 298–353 *in* C. Birkeland, editor. Life and death of coral reefs. Chapman and Hall, N.Y.

Pennings, S.C. 1997. Indirect interactions on coral reefs. Pages 249–272 *in* C. Birkeland, editor, Life and death of coral reefs. Chapman and Hall, N.Y.

Podestá, G.P., and P.W. Glynn. 2001. The 1997–98 El Niño event in Panama and Galápagos: an update of thermal stress indices relative to coral bleaching. Bulletin of Marine Science **69**:43–59.

Reaka-Kudla, M.L. 1997. The global biodiversity of coral reefs: a comparison with rain forests. Pages 83–108 *in* M.L. Reaka-Kudla, D.E. Wilson, and E.O. Wilson, editors. Biodiversity II: Understanding and Protecting our Biological Resources. Joseph Henry Press, Washington, D.C.

Reyes Bonilla, H. 2002. Checklist of valid names and synonyms of stony corals (Anthozoa: Scleractinia) from the eastern Pacific. Journal of Natural History **36**:1–13.

Reyes Bonilla, H., and A. López Pérez. 1998. Biogeography of stony corals (Scleractinia) of the Mexican Pacific. Ciencias Marinas **24**:211–224.

Robertson, D.R. 1998. Do coral-reef fish faunas have a distinctive taxonomic structure? Coral Reefs **17**:179–186.

Strub, P.T., J.M. Mesías, V. Montecino, J. Rutllant, and S. Salinas. 1998. Coastal ocean circulation off western South America. Pages 273–313 *in* A.R. Robinson and K.H. Brink, editors, The sea, the global coastal ocean: regional studies and syntheses, Vol. 11. Wiley, N.Y.

Veron, J.E.N. 1995. Corals in space and time: the biogeography and evolution of the Scleractinia. Comstock/Cornell, Ithaca and London.

Veron, J.E.N. 2000. Corals of the world. Vol. 3, Australian Institute of Marine Science.

Wellington, G.M. 1980. Reversal of digestive interactions between Pacific reef corals: mediation by sweeper tentacles. Oecologia **47**:340–343.

Wellington, G.M. 1982. Depth zonation of corals in the Gulf of Panama: control and facilitation by resident reef fishes. Ecological Monographs **52**:223–241.

Wellington, G.M. 1997. Field guide to the corals and coral reefs of the Galápagos Islands, Ecuador. Proceedings of the Eighth International Coral Reef Symposium **1**:185–202.

Wellington, G.M., and P.W. Glynn. 2007. Coral reef responses to El Niño-Southern Oscillation sea warming events. Pages 342–385 *in* R.B. Aronson, editor, Geological approaches to coral reef ecology: placing the current crisis in historical context. Springer Verlag, N.Y.

Wilkinson, C., editor. 2000. Status of coral reefs of the world: 2000. Global coral reef monitoring network. Australian Institute of Marine Science, Cape Ferguson, Queensland, Australia.

9

Conclusions: An Ecosystem Perspective of Shallow Marine Reefs

Tim R. McClanahan and George M. Branch

Food-web ecology has had a long and somewhat checkered past as it has gone through different stages of development, understanding, and sophistication (chapter 1). For example, early discovery of scale-invariant properties of food webs turned out to be an artifact arising from the level of resolution of food webs that investigators created for heuristic simplicity, and were not characteristic of fully resolved food webs (Hall and Raffaelli 1993). This produced some cynicism about the value of food webs and associated theory, as well as the ability to find general principles of their organization (Polis 1991)—but it also spurred a renaissance in badly needed experimentation that dominates ecology today. Recently, however, the limits to experimental manipulations are returning ecologists to a combination of experimental reductionism and large-scale comparative studies. This book provides a synthesis of marine-reef food-web study where authors have reviewed both the experimental and large-scale patterns as summarized below.

The Structure of Marine Reef Ecosystem

Comparative large-scale studies have yielded increasingly fascinating patterns that when melded with local experiments allow exploration of the extent to which experimental results can be extrapolated over geographic scales (Menge et al. 2003, 2004, Nielsen and Navarrete 2004, Wieters 2005, Navarrete et al. 2005). Recognition of the limits of experiments in large and complex ecosystems has led to an appreciation of the need to identify strong interactions and to understand how human or environmental stresses influence them. Finding strong interactions, and modeling and experimenting with these interactions, hold great promise for an ecological holism approach that will lead to better management of ecosystems (Paine 1980, Berlow et al. 1999). The difficulty is in reducing complex food webs to key strong interactions, as this can be troubled by a limited view of the ecosystem and the subjectivity of investigators.

Paine (1980) has distinguished three approaches to developing food webs. The first has been the description of "connectedness webs," in which all trophic connections between all species are identified. The second concerns "energy-flow webs," in which the magnitude of flows between species is quantified. This is a step forward in identifying species that play dominant roles as either producers or consumers. However, experimental manipulations often reveal that the relative importance of species in community dynamics is not necessarily proportional to their contribution to energy flow. Some have a disproportional influence on other species that is recognized in Paine's third approach, which quantifies interaction strength among species in "functional webs" (Raffaelli 2000).

The notion of keystone species is intimately associated with the interaction-strength concept. First introduced by Paine (1966), the concept initially related to the capacity of predators to restrict species that monopolize primary space and permit their coexistence with less competitive species, thus increasing diversity. The idea has grown to incorporate competitors and mutualist species that are seen to have an "important" effect on other members of the community. Because of the risk that "importance" is simply a measure of the beholder's eye, defining keystone species has been elusive (Bond 2001, chapter 1 this volume). To counter this and in an attempt to introduce more rigor, keystone species have been defined as those that have greater effects on other species than would be expected from their abundance (usually expressed as biomass) (Power et al. 1996). In food-web theory, interaction strength is defined as the sum of the effects of a given species on all "affected" species, whereas the Keystone Index is this measure divided by the relative biomass of that species (chapter 6). Both measures are useful. From a manager's point of view the absolute magnitude of the effects of a species allows identification of species with a powerful influence on the ecosystem; but this becomes particularly critical if the effects are large relative to the species' abundance.

Most ecosystems are composed of many weak and a few strong interactions and there is, therefore, the potential to simplify the complexity of food webs and to discover key interactions and species, which would greatly simplify management (Walker 1992). There are experimental, theoretical, and mathematical ways to explore the identity of strong interactors or keystone species (chapter 6), but in chapter 1, Menge notes that there may be no assured way to identify keystones from prior theoretical knowledge. Theory and calculations may reduce the number of possibilities and point in the right direction, but experience and planned experiments or unplanned, large-scale "natural" experiments that focus on key interactions may prove the best way forward.

Food webs are by definition intrinsically limited to flows of energy and materials involving trophic interactions, but nontrophic facilitation or antagonism among species and environmental forcing can alter community dynamics profoundly, both directly and indirectly. For example, chapter 3 on South African kelp beds describes how the sweeping action of blades of the kelp *Laminaria pallida* clears away grazers. This mechanism is nontrophic, but has the trophic consequences of permitting establishment of kelp sporophytes and expansion of kelp clumps (Velimirov and Griffiths 1979). In a similar vein, Wieters (2005) has shown that in areas of the Chilean coast where upwelling is sustained and nutrient inputs relatively high, algal turfs grow taller and inhibit mussel recruitment. The mechanism is unknown but unlikely to be trophic; but the effects on higher trophic levels requiring mussels as a source of food are obvious and clearly trophic in nature. Bottom-up effects such as these can therefore have strong effects on food webs even though they may involve nontrophic steps that need to be incorporated in addition to food transfers to gain a full understanding of the mechanisms involved in functional webs.

Top-Down Versus Bottom-Up Effects

Both the "bottom-up" effects of productivity and recruitment, and the "top-down" influences of consumption by predators and grazers potentially affect community dynamics. Understanding the relative roles of bottom-up and top-down effects and the circumstances that influence them has become an important focus of food-web studies. Experiments, meta-analyses, and models suggest that they interact strongly in influencing diversity and ecosystem function (Worm et al. 2002). It is expected that diversity will change the number of food web links and that diversity, stability, and productivity will interact in ways that can either enhance or negate the effect of each other (Worm and Duffy 2003). For example, high predation at low levels of productivity can cause low diversity and extinction while at high levels of productivity it can reduce competitive exclusion and maintain high diversity (Worm et al. 2002). As an experimental test of this, Wootton and colleagues (1996) manipulated nutrient supplies and the effects of macro-grazers and showed that removal of the latter powerfully influenced the abundance of algae, whereas enhancement of nutrients had relatively minor effects, and then only in El-Niño years and seasons when nutrients were in short supply. Vinueza and colleagues (2006) demonstrated strong effects of grazers on algal abundance in Galápagos during non-El Niño periods, but none during El Niño years when nutrient availability and high temperatures curtailed algal growth to the point that grazer effects disappeared. Circumstances, therefore, alter the magnitude and consequences of nutrient and grazing effects.

Top-down trophic effects are themselves subject to unpredictable switches in behavior. Chapter 2 on the Aleutian Island reefs illustrates how transfer of attention of the killer whale or *Orca* from oceanic prey to sea otters has reversed the trends brought about by the previous recovery of sea otters, which caused depletion of grazers and proliferation of algae (Estes et al. 1998). This example is particularly salutatory because whereas the effects of hunting sea otters might have been forecast, those associated with the switch in behavior by *Orca* could not have been anticipated. Consequently, the experience of ecologists who have worked and experimented with ecosystems and observed and monitored disturbances or natural experiments is required to develop an intuitive understanding of what constitutes strong interactions in food webs. Clearly, as knowledge develops, the interaction web will change, as was seen in the case of the Aleutian food web with the inclusion of the *Orca*. This process is the thread that will lead toward better understanding and management.

Given the call for ecosystem-based management by many ecologists and resource managers (Cochrane et al. 2004, Pikitch et al. 2004), food-web analysis and theory will have to increasingly become part of management tools. Developing a way to ensure this will require the types of research detailed in this book, including the important roles of physico-chemical conditions and biological interactions.

The Environments

The geographic and physico-chemical environments that the authors describe include a number of common themes: productivity, seasonality, depth, currents, waves, tides, upwelling, and climate cycles such as El Niño, Indian Ocean Dipole, and Pacific Decadal Oscillation fluctuations. A quartet of physical factors emerges as being pivotal—light, temperature, nutrients, and water movements—and these differ in their relative importance in different systems and interact in complex ways. In combination, they have two kinds of effects. First, their "normal" levels set limits to the productivity that can be expected. Some ecosystems are intrinsically more productive than others because they share high nutrient and light availability. Second, their fluctuations inject disturbances that operate at three time scales. At the shortest scale, abrupt changes over hours to days provide no opportunity for individuals to acclimate and they are either tolerant or succumb. Low diversity is a consequence of frequent short-term fluctuations. At the scale of changes over months, acclimation is possible, but population growth can only keep pace with the change if the species is short-lived. Finally at interannual scales, species are adapted to prevailing circumstances but can be catastrophically struck by periodic changes that are beyond their adaptive limits. Some ecosystems are highly dynamic; others relatively stable. Nevertheless, all are prone to periodic catastrophic change. Examples spanning the latitudes are rare flooding events on the west coast of South Africa that eliminate grazers and cause uncontrolled proliferations of opportunistic algae (Branch et al. 1990), and ENSO events that diminish nutrients and increase temperatures and wave forces, with a collapse of primary production and ensuing ripple effects through the food web (Ebeling et al. 1985, Ware and Thompson 2000, Vinueza et al. 2006).

Intra-annual Variability

Most of the ecosystems described experience seasonal changes in currents and upwelling or downwelling, and their associated effects on nutrient delivery. Even in East Africa, where upwelling is limited to the northern region, there is considerable seasonality in the physico-chemical environment. Indeed, seasonal patterns must be an important factor in all marine benthic ecosystems and something to which species are adapted and around which ecosystems are organized. Seasonal effects appear to be most influential on primary producers, both benthic and planktonic, but their effects attenuate while moving up the trophic levels of the food web. This phenomenon is particularly marked in systems such as the southern Benguela where upwelling is pulsed. Under these conditions, phytoplankton may respond quickly to upwelling, but higher trophic levels cannot. Their longer life-spans preclude population-level responses at short time scales, creating a mismatch so that they never realize their full potential before the phytoplankton begins to decline again. This diminishes the potential flow up the food web and restricts and stabilizes stocks higher up the food chain (Hutchings 1992, Field and Shillington 2006). The rate of change of environmental conditions can thus have powerful effects on transfers of energy, potentially limiting food-chain length.

Because reproduction and recruitment are seasonal, feeding and associated nutrition must also have strong seasonal components, but none of the authors described these as being important at the level of populations and biomass of higher trophic levels. This is likely due to the time scale of the life spans, where primary producers have sufficiently short life spans for their populations to respond seasonally, but most consumers do not show population-level seasonal responses. Consumers that are migratory are likely to change seasonally as they shift feeding grounds. There may be some short-lived consumers, such as zooplankton and small benthic worms and crustaceans that can change on a seasonal time scale, but perhaps they are not yet well-enough studied to have been registered as noteworthy, as no such seasonal swings of consumers were noted in the interaction webs created by the various authors. This suggests some flexibility in consumer diets on a seasonal basis to allow them to persist during periods of low food availability and to accommodate shifts in the nature of food resources. For example, Seiderer and Robb (1986) have shown that temperature-adjusted activity of bacteriolytic enzymes in mussels allows them to capitalize on bacteria-laden kelp particles at times when seasonal upwelling exports most phytoplankton-derived particulate food.

Interannual Variability

Interannual fluctuations, such as El Niño-Southern Oscillation (ENSO) fluctuations influence most of the ecosystems, although the Indian Ocean Dipole (IOD) is specific to the Indian Ocean and the Pacific Decadal Oscillation (PDO) specific to the northern Pacific. At this scale, spanning three years to decades, consumers higher in the food web are likely to be most strongly influenced. Examples from the coral-reef food webs in the eastern Pacific and East Africa include coral mortality events that oscillate with ENSO and IOD events. Consumers such as iguanas, cormorants, sea lions, and sharks suffer during ENSO events in the Galápagos, either starving or emigrating (Vinueza et al. 2006). Upwelling and the scale of ENSO and PDO events also influence temperate ecosystems of the Pacific, affecting the degree to which planktonic production and associated ecosystems persist and are changed by the nutrient status. Kelp forests in the Pacific are reduced by warm ENSO events that fluctuate with these oscillations and this is expected to affect the proportion of kelp-dependent consumers in these ecosystems, such as abalone species (Dayton et al. 1999, Vilchis et al. 2005). Consequently, interannual fluctuations will change the structure of food webs on the scale of these oscillations. The intensity of these oscillations is likely to increase with climate warming, and will nuance and complicate efforts at management of these ecosystems.

Latitudinal Variability

Tropical systems tend to have abundant light but to be nutrient-limited, whereas temperate systems have greater supplies of nutrients, but the length of light periods, their intensity, turbidity, and biomass often limit light conversion into gross production. In general, tropical systems have higher gross production and are more diverse than temperate systems but have lower net production. Comparing the physical forces of water motion and light on the same energy basis indicates that in equatorial East Africa light is the largest force, but water motion is significant and might be expected to be a nearly equal force in temperate locations where light is reduced and water motion increased. Seasonality affects all studied ecosystems, regardless of their latitude. In temperate locations there are the obvious winter–summer oscillations but tropical locations are also greatly influenced by the Intertropical Convergence Zone and the less extreme oscillations of its annual migration.

Both temperate and tropical oscillations influence wind and current directions and strength and these influence the upwelling or downwelling conditions that ecosystems experience. The persistence and dominance of kelp forests are influenced by the strength of these factors. South African kelp forests decline as turbidity increases associated with upwelling and ecosystems become more dominated by consumer filter feeders such as mussels and their predators, rock lobsters (Field and Griffiths 1991). Warm ENSOs in the Pacific, however, reduce upwelling and this leads to warm, stable, and nutrient-depleted water that also kills kelp (Dayton et al. 1999). Kelp forests in Chile shrink in areas influenced by warm conditions and consumers change their diets from kelp-associated benthic prey to species more dependent on the water-column (chapter 4).

Although idiosyncrasies exist in all ecosystems, there are five general patterns in physical conditions that seem related to latitude, at least across the spectrum from tropical to temperate latitudes covered in the chapters (fig. 9.1a). Three are well documented: low latitudes are associated with nutrient limitation, high light levels, and high temperatures. Less clear-cut are trends for wave action and short-term fluctuations, which generally increase with latitude. The frequency and extent of short-term fluctuations of these variables are particularly important because they impose environmental stresses. Upwelling, for example, can inflict temperature swings of as much as 10°C in a day, and fluctuations of nutrient and light availability can yield boom-and-bust phytoplankton blooms that decay with resultant oxygen depletion and mass mortalities (Cockcroft 2001). Variability of physical conditions is thus an important element of stress.

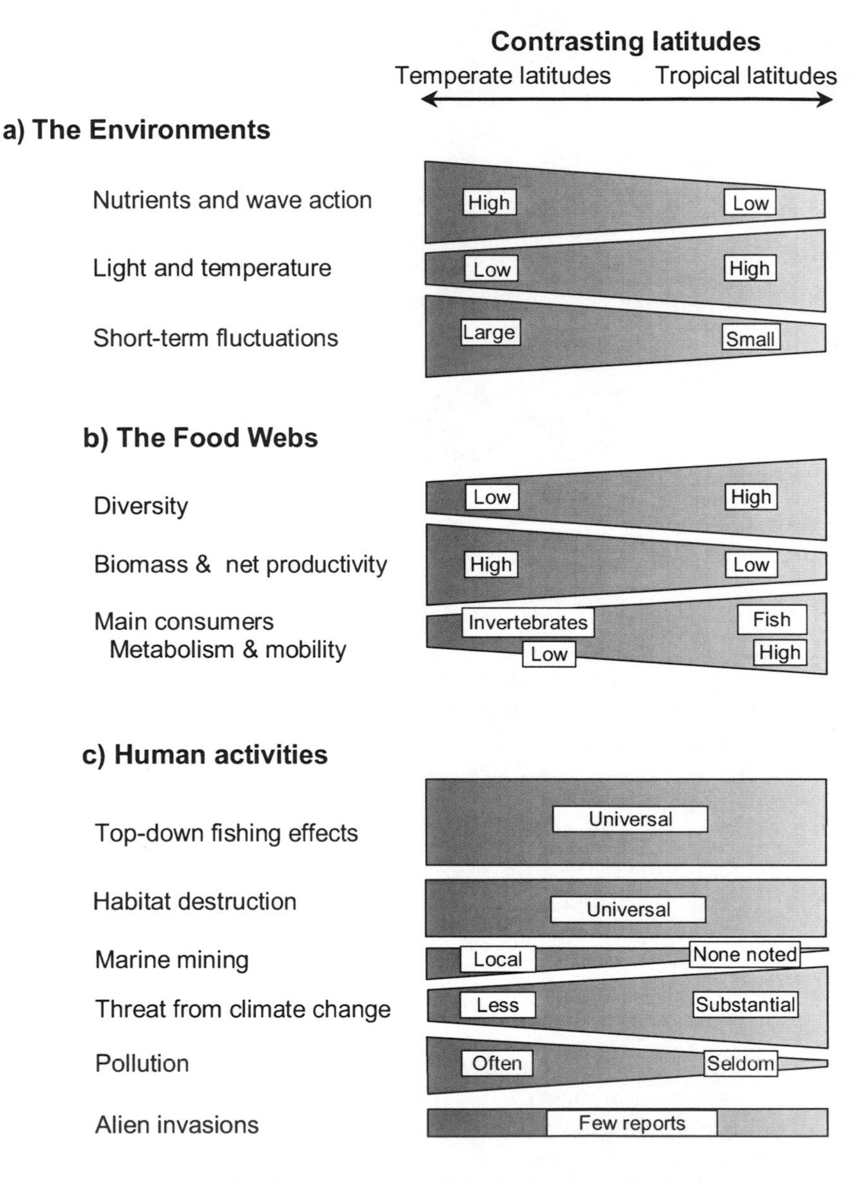

FIGURE 9.1 Summary of (a) the environments; (b) the food webs; and (c) the human activities influencing food webs in temperate and tropical latitudes. The contrasts are deliberately polarized to illustrate general trends, but there are exceptions to these patterns that are discussed in the text.

The Food Webs

General Structure

Geography influences food webs in two important ways: (1) by determining the biogeography and the historical forces of speciation and extinction, it affects the local and regional pools of species that are present, and (2) it influences the physico-chemical setting.

The first factor determines the kinds of species interactions and the complexity of food webs, both of which were listed by all authors as being important aspects of the organization of food webs. Locations such as the Aleutian Islands were only recently reinvaded by temperate species after glaciations and this has resulted in an impoverished ecosystem, with strong but limited numbers of interactions (chapter 2). The physically isolated Galápagos Islands receive inputs from various biogeographic regions and this leads to a mixture of high endemism and cosmopolitan taxa (chapter 6). The pool of tropical and temperate taxa that exists there appears to undergo extinctions and recolonizations as the environment fluctuates over ENSO to decadal and longer time scales.

The second factor shapes the way that water temperature and chemistry, light, and water motion influence and deliver nutrients, organic matter and propagules, thus influencing the productivity and structure of ecosystems (fig. 9.1b). Some food webs, such as the Galápagos and western South Africa are heavily subsidized by oceanic ecosystems and the productivity and biomass of these food webs could not be sustained without significant exogenous inputs. Coral reefs in the downwelling system of East Africa are probably much less subsidized, and the contribution of the oceanic system was discounted when simulating the proposed strong-interaction web (chapter 7). Strong subsidies are likely to affect the importance of interactions and create alternate pathways that are expected to influence stability, production, and resource-extraction potential (Polis et al. 1997). Even in downwelling systems such as East Africa, the high biomass of herbivores cannot be entirely explained by in situ production, and may be subsidized by adjacent seagrass beds that provide energy inputs in the form of detritus.

The eastern Pacific and Atlantic illustrate the importance of interactions between biogeographic history and local physico-chemical factors. Here, upwelling and downwelling can interact in interesting ways and local factors such as the existence of capes, volcanic islands, or mountain ranges can reduce wind forces and associated upwelling and create locally more stable conditions. This creates some of the patchiness in time and space evident in ecosystems (chapters 3–5 and 7). In the tropical eastern Pacific a combination of isolation and highly variable environment (mixing of cold and warm waters, unpredictable upwelling, and periodic ENSO events) has produced a low-diversity coral-reef ecosystem. The warmer but more constant downwelling Caribbean has a higher diversity despite its isolation from the high diversity of the Indo-Pacific. This has influenced the food-web structure: despite low levels of diversity in the basal (coral) species, there is considerable diversity in the consumers of corals.

This is also the case in East African coral reefs, where the highest diversity of fish is found at intermediate trophic levels, between levels 2.5 and 3.5. Consequently, biodiversity within ecosystems is not clearly proportional to the energy stock or flow, but appears to be influenced by history, species interactions, and environmental conditions.

Ecosystem Components

Primary production and microbes Primary producers on shallow-water reefs are derived from a few main functional groups, namely kelp, turf algae, coralline algae, red and brown macrophytic algae and corals with their endosymbiotic algae, together with a through-put of phytoplankton, detritus, and particulate organic matter. The proportions of these groups change in temperate and tropical systems with corals becoming less abundant in temperate areas and in stressed tropical environments, as seen in the eastern Pacific and in East Africa after anomalous temperatures. In these systems, brown frondose algae are most often canopy-forming, with the strata beneath them created by red frondose algae, and a mix of red coralline and turf algae as the most basal taxa. Kelp assumes major importance in cool-temperate upwelled systems, but is absent from subtropical and tropical areas where low nutrients and abundant herbivorous fishes are likely to restrict its abundance (Gaines and Lubchenco 1982, McClanahan et al. 2003, Floeter et al. 2005). Phytoplankton is more important in temperate upwelled systems than elsewhere but, even there, its standing stock is low and its importance is only realized when account is taken of its productivity and transfer rates over reefs: because reefs are stationary but phytoplankton flows over them, currents, tides, coastal topography, and waves act as gateways for replenishing supplies (Field and Griffiths 1991, Bustamante and Branch 1996).

In most of the food webs examined, the role of bacteria and microheterotrophs is poorly understood and seldom explicitly studied. In cases where bacterial flows have been quantified, they account for about 10% of the energy flow to primary consumers and are particularly important in terms of nitrogen enrichment of particulate organic matter (Azam et al. 1983).

Consumers Consumers in reef food webs often depend on one of two forms of primary production: living plant matter or detritus, with bacteria transforming and adding to the value of the latter. Both sources are listed as important in all of the ecosystems described but they may vary in their proportion of use. For example, some herbivores and particulate feeders in Alaskan, South African, and Chilean kelp forests appear to capture and largely depend on kelp drift and particulate matter as their source of primary production

(Duggins and Eckman 1994, Bustamante et al. 1995, Bustamante and Branch 1996, Rodriguez 2003). Whether herbivores rely mainly on living plants or drift material profoundly influences their role. In South African kelp beds sea urchins rely largely on drift kelp, and experimental removal of urchins has little effect on the stocks of living macroalgae (Day and Branch 2002). In kelp forests such as those in California, there are significant levels of herbivory on living plants, mostly attributable to sea urchins, and their influence on algal stocks there is substantial. Switches between these two modes of feeding may take place, depending on the availability of drift kelp (Harrold and Pearse 1987).

There is a general trend for fish to play a dominant role as herbivores at low latitudes but to be replaced by invertebrates at higher latitudes. The abundance of Chilean fish herbivores, as on many coastlines (Floeter et al. 2005), depends on latitude, with no herbivorous fishes poleward of 33°S, increasing to nearly one quarter of the herbivore biomass at 29°S. The relative contribution made by fish and invertebrates has important consequences. Compared to most invertebrates, fish are mobile, active, and have high metabolic rates. They can cover wide areas, move on a short-term basis depending on conditions, undertake large-scale migrations, and have greater food requirements. Both their mobility and their metabolic needs influence their effects on ecosystems. Fish are by far the dominant herbivores/detritivores in unfished coral reefs of East Africa and elsewhere, but this role is reversed and sea urchins occupy this trophic level in heavily fished reefs where both the predators and competitors of sea urchins are reduced, similar to the situation in the Caribbean before the massive die-off of the dominant sea urchin, *Diadema antillarum*, in 1983 (Hay 1984, Lessios 2005). Consequently, sea urchins can occupy a dominant role as herbivores in tropical ecosystems if predation is sufficiently low to allow their persistence.

Herbivore trophic dynamics may also be influenced by the chemical defenses of algae. For example, phlorotannins are an order of magnitude greater in southern than northern hemisphere kelps and rockweeds of the Pacific (Steinberg 1992, Steinberg et al. 1995) and high concentrations of polyphenols have been reported in South African kelps of the Benguela (Tugwell and Branch 1989). The chemical defenses of living plants may increase the role of kelp drift and detritus in these ecosystems. The inclusion of the otter in the north Pacific kelp-forest food web and its high demand on benthic invertebrate grazers, mainly sea urchins, reduces herbivory on kelp and is therefore the most likely explanation for the lower chemical defenses (chapter 2; Steinberg et al. 1995).

Trophic structure and cascades Most food chains are short, with three to four links (Raffaelli 2000). Environmental stress models (Menge and Sutherland 1987) propose shorter chain lengths in stressed environments, and productivity models (Fretwell 1987) assume a relationship between productivity and chain length. However, all the ecosystems described in this book had a total of 3.5 to 4.5 trophic levels, with no evidence that tropical systems have more or less levels than temperate food webs, or that productivity is related to the number of trophic levels. This may reflect the relatively constant conditions associated with subtidal reefs, which span a relatively small part of the stress and productivity gradients that exist across different ecosystems, and suggests dynamic limits to benthic reef systems and the possibility for some general principles in their organization (Pimm 1991).

Most theoretical models of ecosystems are based on a three-trophic-level simplification (chapter 1), but in reality this is not the norm for shallow marine ecosystems. This is likely to influence the types of predator–prey dynamics and the types of trophic cascades that might be expected, depending on how permanent the higher trophic levels are and how influential they are in determining abundance in lower trophic levels.

The kelp–urchin–otter, kelp–grazer–lobster, and the algae/coral–urchin/herbivorous fish–triggerfish interaction chains described for the Aleutian Islands, South Africa, and East Africa respectively fit well with the three-trophic level model of ecosystems (chapters 2, 3, and 7), although the depletion of large reef-fish by overfishing may have artificially shortened the chain in South Africa and East Africa (McClanahan 1995, Griffiths 2000). The fourth trophic level in East Africa constitutes a variety of piscivores that are often positively associated with triggerfish because they are all influenced by fishing. The Aleutian model is nuanced by the inclusion of the *Orca* as a possible fourth trophic level. The degree to which *Orca* is, in fact, a permanent or transient member of this food web and the influence of reductions in their offshore whale and fish resources on their behavior will affect the likely dynamics of this food web in the absence of human influences on offshore fisheries.

In theory, a three-level food web should result in the suppression of herbivores and release primary producers, whereas a four-level web will move these effects one level up, with top predators controlling secondary consumers, releasing herbivores that then restrict primary producers. Similarly, in a two-level web primary producers will be scarce because herbivores are abundant in the absence of predators (Wootton and Power 1993). In South African kelp beds the three-level is demonstrated in areas with high rock-lobster abundance are synonymous with a proliferation of algae; but in areas where large rock lobsters are scarce or absent because of either fishing or environmental circumstances, the system effectively becomes a two-level one in which algae are scarce, conforming to the predictions for a two-level web (Branch et al. 1987, Barkai and Branch 1988, Mayfield and Branch 2000). Detection of four-step sequences with the expected outcome has been more elusive. There is, for example, no evidence that the fourth trophic level reduces the third level, resulting in a prey release of the second level in East African marine parks that have been studied (McClanahan et al. 1999). Introduction of *Orca* into the fourth trophic level of the Aleutian food web has, however, transformed that system from an algal-dominated three-level web into an algal-depauperate four-level system, reversing the previous control of herbivorous urchins by otters (Estes et al. 1998). The Aleutians therefore "behave" as predicted by the theoretical expectations of three- and four-level food webs.

Why would a number of the ecosystems behave like a three-chain model when additional trophic levels and other pathways exist? One possibility is that there is some body-size refuge for fish where above a certain size their predators are ineffective and unable to reduce numbers of the larger and therefore long-lived prey. This has been shown for herbivorous fishes in coral reefs (Mumby et al. 2006) and may be the case for the triggerfish found in East Africa. Additionally, it is possible that high metabolism and associated high consumption rates override trophic-chain dynamics, which may explain why the otter and triggerfish cause the demise of sea urchins. High metabolism is associated with a need for a protein-based diet at the third trophic level or above, and generally organisms with high metabolic rates feed on organisms with low metabolic rates and reduce them to low levels. Categorization of taxa into trophic levels may be less critical to understanding dynamics than a consideration of their metabolic rates and productivity (McClanahan 1992).

Given that (1) most ecosystems presented here are webs and not chains; (2) many species occupy intermediate rather than discrete trophic levels; (3) species change their diets as they develop and age; and (4) diets are flexible, trophic dynamics will be considerably more complicated than those predicted by simple models and theory. Omnivory can moderate the intensity of direct predator–prey interactions and has, therefore, been seen as a factor that reduces trophic cascade effects (Menge et al. 1986, Polis and Strong 1996). The important role of detritus also complicates the predictions of simple models as it creates an alternate pathway for production and throughput, and detritivores draw on a pool of material derived from various trophic levels and enriched by bacteria. In cases where it has been quantified, more than 50% of the primary production passes through the detrital path (chapters 3 and 6), much of it via particulate matter consumed by suspension feeders. In this respect, marine benthic food webs differ radically from terrestrial ones, for which there is no equivalent feeding group (Denny 1990).

Disturbance, stress, and fishing often have their greatest effects on higher trophic levels, releasing prey so that their abundances can increase significantly. The degree to which consumers affect their prey also seems linked to the position they occupy in the food chain. In chapter 6, on the Galápagos ecosystem, the "keystoneness" of species (their effect relative to their abundance) rose progressively through the trophic levels and was concentrated in the upper levels, although the strength of interactions among species was uniform among levels. Both phenomena seem general properties of marine benthic food webs.

Patterns of species composition Intriguing similarities in the taxa and their position in the food webs of reef ecosystems from disparate areas of the world suggest the potential for basic principles of food-web organization based on trophic level, taxa, functional groups, and disturbance or environmental stress. For example, fish, sharks, birds, and mammals often occupy the highest trophic levels above three and attributes of their size, mobility, and energy requirements must play an important role in determining the role of species in this trophic position. Invertebrates such as snails, whelks, and crustaceans are often dominant at trophic level 3 in temperate ecosystems, and can be sufficiently abundant to exert a control on the level below, such as the effects of rock lobsters in the Benguela ecosystem (Mayfield and Branch 2000).

In unfished and unstressed tropical ecosystems, mobile invertebrates are a small part of the biomass, although they become more important in areas that are fished or stressed. Fishes dominate most of the consumer niches in the tropics, but many feed and rely on invertebrates and it may be that the high biomass of fish is maintained by a high net production of invertebrates with a low biomass but high turnover. This is similar to what is found for turf-forming algae in coral reefs, where the high net production and turnover of these algae maintain a consumer biomass that is equal to or greater than the primary producer biomass (McClanahan 1992). As a general principle, assessments of the role of species based on biomass will underestimate the importance of small and productive species with a high turnover. As examples, Shafir and Field (1980) showed that in South African kelp beds small predatory isopods accounted for only 10% of the predator biomass but > 50% of the consumption and production, and Barkai and Branch (1988) demonstrated that settling barnacles, although so trivial in biomass as to be neglected in surveys, at times contribute the bulk of the diet of rock lobsters because of their high turnover.

Temperate grazers tend to be sea urchins, whereas fish are the predominant grazers in the tropics, but both are found in each system, the relative proportions changing with latitude. Small benthic crustaceans, bivalves, worms, and sponges feed on detritus in most systems but many fishes are also important consumers of detritus in tropical ecosystems. There are some general principles of organization of taxa that appear to depend on the degree of stress or disturbance, which can in turn be related to latitude, fishing, or environmental perturbations. Fish tend to dominate most consumer positions in the absence of human disturbances or environmental stresses associated with latitude. Fishing, irrespective of latitude, appears to be the key disturbance that reduces their dominance, leading to replacement by invertebrates.

Six general characteristics of the structure of food webs thus emerge (fig. 9.1b). At low latitudes, (1) diversity is high but (2) biomass and (3) net productivity are low. In terms of taxa (4), the tropics are dominated by fish, whereas invertebrates play stronger roles in temperate areas. Associated with this, low latitudes seem predominated by (5) active, mobile species with (6) relatively high metabolic rates, but temperate climes have a greater prevalence of sedentary species with low metabolic rates. Both metabolic demands and relative mobility will powerfully influence the way in which food webs operate.

Human Influences

Previous reviews of current and projected future threats to intertidal and subtidal reefs have singled out a suite of six threats of primary importance: fishing, habitat destruction, pollution, climate change, marine mining, and alien invasions (Thompson et al. 2002, Branch et al. 2008, Polunin 2008). The contributors to this book list a variety of possible human influences that range from local scales such as mining for diamonds in South Africa to global scales of climate change that may be increasing the intensity of oceanographic oscillations (fig. 9.1c).

Primary Effects

Most authors focus on the direct and indirect effects of fishing, partly because of its huge capacity to alter food webs (Dayton et al. 1995, Steneck 1998, McManus et al. 2000, Halpern et al. 2006). Without exception, the authors of the various chapters agree that fishing and its attendant effects constitute the greatest and most universal threat. Fisheries differ in scale around the world, with multi-species, often artisanal, fisheries prevalent in the tropics but more focused single-species industrial operations the norm in temperate climes. This immediately has important consequences for management styles and for the effects of fishing on food webs. Single-species fishing, normally aimed at abundant and high-value species, focuses on a single level of the food web and its effects depend on which level is targeted, but multi-species harvesting spans different trophic levels. There are examples of fishing down (Pauly et al. 1998) and more of fishing through the food webs (Essington et al. 2006). Fishing down the food web can often result in trophic cascades but the effects of fishing through the food web are poorly studied and expected to affect food webs in currently unappreciated ways.

All contributors to this book record top-down effects of fishing/harvesting that ripple through food webs, yielding indirect effects on lower trophic levels. The near extirpation of otters from the Aleutians, reductions of predatory fish in East Africa, and depletions of large rock lobsters in the Benguela are classic cases. In South Africa, rock lobster mandibles found in shell middens are twice the size of those of currently harvested individuals. These influences are by no means all recent, as seen in the North Pacific where the Steller's sea cow was driven to extinction by colonial Europeans, and before that, aboriginal hunters achieved serial depletion of resources in California (chapters 2 and 5). More recently, major reductions in whale populations in the nineteenth and early twentieth centuries may still be influencing Aleutian kelp forests today. In the more remote islands of the Galápagos, the resources have been serially depleted since the 1940s, including the near-elimination of sea cucumbers (Hearn et al. 2005), and fishing is expanding to increased reliance on the larger pelagic systems that surround the islands.

In some cases, fishing starts by depleting top trophic levels and moves progressively to lower levels as resources collapse—"fishing down the food web" (Pauly et al. 1998). However, if harvesting is directed at intermediate trophic levels, for example pelagic filter-feeding fish, it can have a "wasp-waist" effect, diminishing both food supplies to higher trophic organisms such as piscivorous seabirds and consumption of organisms at lower trophic levels (Cury et al. 2000). Many large fisheries harvest from multiple trophic levels but their management is based on a series of single-species assessments that is often poorly integrated into a vision of the food web and ecosystem effects (Essington et al. 2006).

In a number of ecosystems, including coral reefs, depletion of higher trophic levels has produced a simple algal–sea urchin system, but the market for sea urchins in Asia now has the potential to reduce the system to an empty algal-dominated forest (Redford and Feinsinger 2001, Berkes et al. 2006). And in what could be the final stage in serial depletion, kelp itself is being harvested for production of agar and other industrial compounds, and to supply abalone farms with feed (Troell et al. 2006). Apart from the direct effects of fishing on target species and the indirect cascade effects on other members of the community, fishing also has the capacity for habitat destruction, as for example corollary damage to coral ecosystems by dynamite or cyanide fishing, or the impacts of dredging on benthic communities (Steneck 1998, Jennings and Kaiser 1998, McManus et al. 2000, Koslow et al. 2001, Jackson et al. 2001).

Secondary Effects

Most of the authors of the chapters also consider global climate change to be a central threat to ecosystems, most particularly in the tropics where coral reefs stand poised on the limits of their thermal tolerances and coral bleaching is already an existent problem (chapters 7 and 8). Galápagos, which sits astride three major ocean currents, is especially vulnerable to climate change, as illustrated by the dramatic changes to food webs that take place during El Niño events (Glynn 1990, Vinueza et al. 2006). Kelp forests are also often decimated by strong El Niño events that are expected to become stronger as baseline temperatures rise (Dayton et al. 1999).

Pollution is now universal, although its effects are concentrated around industrial and shipping activities in temperate areas. Point-source pollution has the capacity to devastate ecosystems, and often depletes top predators and grazers first, leaving opportunistic algae as dominants. But by definition its effects are local. More insidious and of greater significance are the wide-scale effects of terrestrial nutrient and chemically loaded runoff that lead to eutrophication and anoxic "dead zones" over areas as large as 15,000 km^2 in the Gulf of Mexico (Turner and Rabalais 1994). Mining has similarly destructive effects on habitats. Although thus-far local in scale, impending mining plans threaten wider areas.

Alien invasive species have been identified as a major and universal threat to marine ecosystems (Carlton 1999, Ruíz et al. 2000, Grosholz 2002), but subtidal reefs globally appear to have escaped largely unscathed. Apart from the transforming impacts of *Caulerpa taxifolia* in the Mediterranean (Galil 2000) and the devastation of benthic invertebrates in Tasmania and Southern Australia by the northern Pacific seastar *Asterias amurensis* (Ross et al. 2002), few instances of subtidal reefs being substantially affected by invasive aliens have been recorded. Whether this reflects resistance to invasion or failure to detect invasions is an open question. It is possible that the effects are subtle and too ambiguous to be able to pin down alien invasive species as the cause.

The frequencies of harmful algal blooms appear to be on the increase, and arrival of new species may be attributable to worldwide transport of species in ship ballast (Carlton 1999). Toxic species and depletion of oxygen by blooms have the capacity to initiate mass mortalities such as the stranding of 2000 tons of rock lobsters recorded in chapter 3. The magnitude of such events will clearly transform food webs, as evidenced by the release of sea urchins after decimation of rock lobsters by harmful algal blooms. Harris and Tyrrell (2001) caution that many of these human-induced problems act synergistically so that the effects of alien species, climate change, and overfishing may compound one another.

Management Actions

In some cases, the harvesting of one resource influences another, as described for the rock lobster–abalone–sea urchin interactions in South African kelp forests (chapter 3). Such cascades have implications for both fisheries management and conservation, and have brought about the desire for ecosystem management that considers species interactions and trade-offs of various management alternatives. Single-species management has up until now been the backbone of fisheries management, and is likely to remain so for some time because the sophisticated modeling that underpins it has not yet reached a stage where it can be extended to predictive multi-species modeling.

New approaches are required, and three have made huge strides in recent years: co-management, ecosystem-based management, and marine protected areas. Co-management simply means substantive participation and inclusive management that incorporates both managers and users in a joint responsibility for management (Berkes et al. 2001). At its best, it earns ownership, cooperation and responsibility, but it is neither simple to secure nor a panacea for all problems.

An ecosystem approach involves consideration of four components: (1) bycatch; (2) gear effects on the habitat; (3) indirect effects on other species such as predators, prey, competitor, and mutualist species; and (4) potential alteration of environmental conditions and ecological processes. If adopted, managers will no longer be able to focus on fishers and the fish they wish to catch; rather, the total effects of fishing on the ecosystem, including the human element, will need to be incorporated into management procedures. Difficult questions will arise. For example, if fishing threatens a species that is in competition with fishers, which gets precedence?

Marine protected areas or management that closes fisheries in defined areas form a component of ecosystem management and are clearly an important tool as they keep ecosystems intact. They have a proven record of success as a means of allowing resources to recover (Roberts et al. 2001, Halpern 2003, Micheli et al. 2004, McClanahan et al. 2007), and have also been invaluable in providing base-line information on the operation of ecosystems in the absence of fishing; but they do have two drawbacks. First, they are "necessary but not sufficient" (Allison et al. 1998) because they cannot alone provide all the protection needed, especially if fishing around them is maintained at levels that makes them ecologically unsustainable "islands." They also offer no defense against natural disturbances and against pollution such as oil spills. They cannot exclude alien species, and they provide only partial respite against fishing for mobile species that pass through their borders. Second, by definition, they exclude resource users, with particularly tough consequences for poor subsistence fishers for whom local access to fishing is a means of survival rather than an optional income (McClanahan 1999).

Closures need to be designed in a way that their effects minimize the loss of harvesting but still maintain the ecosystem benefits (Halpern and Warner 2002, McClanahan et al. 2006). Outside these marine protected areas, harvesting models need to become more inclusive of multiple species, and to determine the various effects of harvesting on the ecosystem. This will require the food-web approach described here. It can help replace the serial depletion that is the result of our present nonmanagement market-driven system that dominates most of the world's fisheries. There will be a need to increase the sophistication of our understanding of food webs, and the monitoring, decision-making, and control of harvesting, and ultimately to translate this sophistication into simple management heuristics that can be implemented by resource users in collaboration with managers (McClanahan and Castilla 2007). We hope this book takes an important step in the direction toward this goal.

References

Azam, F., T. Fenchel, J.G. Field, J.S. Gray, L.A. Meyer-Reil, and F. Thingstad. 1983. The ecological role of water-column microbes in the sea. Marine Ecology Progress Series **10**:257–263.

Allison, G.W., J. Lubchenco, and M.H. Carr. 1998. Marine reserves are necessary but not sufficient for marine conservation. Ecological Applications **8**(Suppl.):S79–S92.

Barkai, A., and G.M. Branch. 1988. Energy requirements for a dense population of rock lobsters *Jasus lalandii*: novel importance of unorthodox food sources. Marine Ecology Progress Series **5**:83–96.

Berkes, F., R. McMahon, P. McConney, R. Pollnac, and R. Pomeroy. 2001. Managing small-scale fisheries: alternative directions and methods. International Development Research Centre (IDRC), Ottawa, Canada.

Berkes, F., T.P. Hughes, R.S. Steneck, J.A. Wilson, D.R. Bellwood, B. Crona, C. Folke, L.H. Gunderson, H.M. Leslie, J. Norberg, M. Nystrom, P. Olsson, H. Osterblom, M. Scheffer, and B. Worm. 2006. Globalization, roving bandits, and marine resources. Science **311**:1557–1558.

Berlow, E.L., S.A. Navarrete, C.J. Briggs, M.E. Power, and B.A. Menge. 1999. Quantifying variation in the strengths of species interactions. Ecology **80**:2206–2224.

Bond, W. 2001. Keystone species—hunting the snark? Science **292**:63–64.

Branch, G.M., A. Barkai, P.A.R., Hockey, and L. Hutchings. 1987. Biological interactions: causes or effects of variability in the Benguela Ecosystem? South African Journal of Marine Science **5**:425–445.

Branch, G.M., S. Eekhout, and A.L. Bosman. 1990. Short-term effects of the Orange River floods on the intertidal rocky-shore communities of the open coast. Transactions of the Royal Society of South Africa **47**:331–354.

Branch, G.M., R.C. Thompson, T.P. Crowe, J.C. Castilla, O. Langmead and S. J. Hawkins. 2008. Rocky intertidal shores: prognosis for the future. In press *in* Polunin, N.V.C., editor, Aquatic ecosystems: trends and global prospects. Cambridge University Press, Cambridge, U.K.

Bustamante, R.H., and G. M. Branch. 1996. The dependence of intertidal consumers on kelp-derived organic matter on the west coast of South Africa. Journal of Experimental Marine Biology and Ecology **196**:1–28.

Bustamante, R.H., G.M. Branch, and S. Eekhout. 1995. Maintenance of an exceptional grazer biomass in South: subsidy by subtidal kelps. Ecology **76**:2314–2329.

Carlton, J.T., 1999. The scale and ecological consequences of biological invasions in the World's oceans. Pages 195–212 *in* Sandlund, O.T. et al. editors,, Invasive species and biodiversity management. Kluwer, the Netherlands.

Cochrane, K.L., Augustyn, C.J., Cockcroft, A.C., David, J.H.M., Griffiths, M.H., Groeneveld, J.C., Lipinski, M.R., Smale, M.J., Smith, C.D., and R.Q.J. Tarr. 2004. An ecosystem approach to fisheries in the southern Benguela context. African Journal of Marine Science **26**:9–35.

Cockcroft, A.C. 2001. *Jasus lalandii* "walkouts" or mass strandings in South Africa during the 1990s: an overview. Marine and Freshwater Research **52**:1085–1094.

Cury, P., Bakun, A., Crawford, R.J.M., Jarre, A., Quiñones, R.A., Shannon, L.J., and H.M. Verheye. 2000. Small pelagics in upwelling systems: patterns of interactions and structural changes in "wasp-waist" ecosystems. ICES Journal of Marine Science **57**:603–618.

Day, E., and G.M. Branch. 2002. Effects of sea urchins (*Parechinus angulosus*) on juveniles and recruits of abalone (*Haliotis midae*). Ecological Monographs **72**:133–149.

Dayton, P.K., S.F. Thrush, M.T. Agardy, and R.J. Hofman 1995. Environmental effects of marine fishing. Aquatic Conservation: Marine and Freshwater Ecosystems **5**:205–232.

Dayton, P.K., M.J. Tegner, P.B. Edwards, and K.L. Riser, Kristin L. 1999. Temporal and spatial scales of kelp demography: the role of oceanographic climate. Ecological Monographs: **69**:219–250.

Denny, M.W. 1990. Terrestrial versus aquatic biology: the medium and its message. American Zoologist **30**:111–121

Duggins. D.O., and J.E. Eckman. 1994. The role of kelp detritus in the growth of benthic suspension feeders in an understory kelp forest. Journal of Experimental Marine Biology and Ecology **176**:53–68.

Ebeling, A.W., D.R. Laur, and R.J. Rowley. 1985. Severe storm disturbances and reversal of community structure in a southern California kelp forest. Marine Biology **84**:287–294.

Essington, T. E., A. H. Beaudreau, and J. Wiedenmann. 2006. Fishing through marine food webs. Proceedings of the National Academic Sciences, USA **103**:3171–3175.

Estes, J.A., M.T. Tinker, T.M. Williams, and D.F. Doak, 1998. Killer whale predation on sea otters linking oceanic and nearshore ecosystems. Science **282**:473–476

Field, J.G., and C.L. Griffiths. 1991. Littoral and sublittoral ecosystems of southern Africa. Pages 323–346 *in* A.C. Mathieson and P.H. Niehaus, editors. Ecosystems of the world **24**. Elsevier, Amsterdam.

Field, J.G., and F.A. Shillington. 2006. Variability of the Benguela Current System (16,E). Pages 833–861 *in* A.R. Robinson and K.H. Brink, editors, The sea **14**. President and Fellows of Harvard College, Harvard.

Floeter, S.R., M.D. Behrens, C.E.L. Ferreira, M.J. Paddack, and M.H. Horn. 2005. Geographical gradients of marine herbivorous fishes: patterns and processes. Marine Biology **147**:1435–1447.

Fretwell, S.D. 1987. Food chain dynamics: the central theory of ecology? Oikos **50**:291–301.

Gaines, S. D., and J. Lubchenco. 1982. A unified approach to marine plant-herbivore interactions. Annual Review of Ecology and Systematics **13**:111–138.

Galil, B.S. 2000. A sea under siege—alien species in the Mediterranean. Biological Invasions **2**:177–186.

Glynn, P.W., 1990. Coral mortality and disturbances to coral reefs in the tropical eastern Pacific. Pages 55–126 *in* P.W. Glynn, editor, Global ecological consequences of the 1982–83 El Niño-Southern Oscillation. Elsevier, Amsterdam.

Griffiths, M. 2000. Long-term trends in catch and effort of commercial line-fish of South Africa's Cape Province. Snap-shots of the 20th century. South African Journal of Marine Science **22**:81–110.

Grosholz, E. 2002. Ecological and evolutionary consequences of coastal invasions. Trends in Ecology and Evolution **17**:22–27.

Hall, S.J., and D.G. Raffaelli. 1993. Food webs: theory and reality. Advances in Ecological Research **24**:187–239.

Halpern, B.S. 2003. The impact of marine reserves: do marine reserves work and does reserve size matter? Ecological Applications **13**(Suppl.):S117–S137.

Halpern, B.S., and R. Warner. 2002. Matching marine reserve design to reserve objectives. Proceedings of the Royal Society **270**:1871–1878.

Halpern, B.S., K. Cottenie, and B.R. Broitman. 2006. Strong top-down control in southern California kelp forest ecosystems. Science **312**:1230–1232.

Harris, L.G., and M.C. Tyrrell. 2001. Changing community states in the gulf of Maine: synergism between invaders, overfishing and climate change. Biological Invasions **3**:9–21.

Harrold, C., and J.S. Pearse. 1987. The ecological role of echinoderms in kelp forests. Echinoderm Studies **2**:137–233.

Hay, M.E. 1984. Patterns of fish and urchin grazing on Caribbean coral reefs: are previous results typical? Ecology **65**:446–454.

Hearn, A., P. Martínez, M.V. Toral-Granda, J.C. Murillo, and J. Polovina. 2005. Population dynamics of the exploited sea cucumber *Isostichopus fuscus* in the western Galápagos Islands, Ecuador. Fisheries Oceanography **14**:377–385.

Hutchings, L. 1992. Fish harvesting in a variable productive environment—searching for rules or searching for exceptions? South African Journal of Marine Science **12**:297–318.

Jackson, J.C.B., M.X. Kirby, W.H. Berger, K.A. Bjorndal, L.W. Botsford, B.J. Bourque, R.H. Bradbury, R. Cooke, J. Erlandson, J.A. Estes, T.P. Hughes, S. Kidwell, C.B. Lange, H.S. Lenihan, J.M. Pandolfi, C.H. Peterson, R.S. Steneck, M.J. Tegner, and R.R. Warner, 2001. Historical overfishing and the recent collapse of coastal ecosystems. Science **293**:629–638.

Jennings, S., and M.J. Kaiser, 1998. The effects of fishing on marine ecosystems. Advances in Marine Biology **34**:203–314.

Koslow, J.A., K. Gollett-Holmes, J.K. Lowry, T. O'Hara, G.C.B. Poore, and A. Williams. 2001. Seamount benthic macrofauna off southern Tasmania: community structure and impacts of trawling. Marine Ecology Progress Series **213**:111–125.

Lessios, H.A. 2005. *Diadema antillarum* populations in Panama twenty years following mass mortality. Coral Reefs **24**:125–127.

Mayfield, S., and Branch, G.M. 2000. Inter-relationships among rock lobsters, sea urchins and juvenile abalone: implications for community management. Canadian Journal of Fisheries Aquatic Science **57**:2175–2187.

McClanahan, T.R. 1992. Resource utilization, competition and predation: a model and example from coral reef grazers. Ecological Modelling **61**:195–215.

McClanahan, T.R. 1995. A coral reef ecosystem-fisheries model: impacts of fishing intensity and catch selection on reef structure and processes. Ecological Modelling **80**:1–19.

McClanahan, T.R. 1999. Is there a future for coral reef parks in poor tropical countries? Coral Reefs **18**:321–325.

McClanahan, T. R., and J.C. Castilla, editors. 2007. Fisheries management: progress towards sustainability. Blackwell, London.

McClanahan, T. R., N. A. J. Graham, J. M. Calnan, and M. A. MacNeil. 2007. Towards pristine biomass: reef fish recovery in coral reef marine protected areas in Kenya. Ecological Applications **17**:1055–1067.

McClanahan, T.R., N.A. Muthiga, A.T. Kamukuru, H. Machano, and R.W. Kiambo. 1999. The effects of marine parks and fishing on coral reefs of northern Tanzania. Biological Conservation **89**:161–182.

McClanahan, T.R., E. Sala, P. Stickels, B.A. Cokos, A. Baker, C.J. Starger, and S. Jones. 2003. Interaction between nutrients and herbivory in controlling algal communities and coral condition on Glover's Reef, Belize. Marine Ecology Progress Series **261**:135–147.

McClanahan, T.R., E. Verheij, and J. Maina. 2006. Comparing management effectiveness of a marine park and a multiple-use collaborative management area in East Africa. Aquatic Conservation: Marine and Freshwater Ecosystems **16**:147–165.

McManus, J.W., L.A. Menez, K.N. Kesner-Reyes, S.G. Vergara, and M.C. Ablan. 2000. Coral reef fishing and coral-algal phase shifts: implications for global reef status. ICES Journal of Marine Science **57**:572–578.

Menge, B.A., Blanchette, C., Raimondi, P.T., Freidenberg, T., Gaines, S.D., Lubchenco, J., Lohse, D., Hudson, G., Foley, M., and J. Pamplin. 2004. Species interaction strength: Testing model predictions along an upwelling gradient. Ecological Monographs **74**:663–684.

Menge, B.A., J. Lubchenco, L.R. Ashkenas, and F. Ramsey. 1986. Experimental separation of effects of consumers in sessile prey in the low zone of a rocky shore in the Bay of Panama: direct and indirect consequences of food-web complexity. Journal of Experimental Marine Biology and Ecology **100**:225–269.

Menge, B.A., Lubchenco, J., Bracken, M.E.S., Chan, F., Foley, M.M., Freidenberg, T.L., Gaines, S.D., Hudson, G., Krenz, C., Leslie, H., Menge D.N.L., Russell, R., and M.S. Webster. 2003. Coastal oceanography sets the pace of rocky intertidal community dynamics. Proceedings of the National Academy of Sciences USA **100**:12229–12234.

Menge, B.A., and J.P. Sutherland. 1987. Community regulation: variation in disturbance, competition, and predation in relation to environmental stress and recruitment. American Naturalist **130**:730–757.

Micheli, F., B.S. Halpern, L.W. Botsford, and R.R. Warner. 2004. Trajectories and correlates of community change in no-take marine reserves. Ecological Applications **14**:1709–1723.

Mumby P.J., C.P. Dahlgren, A.R. Harborne, C.V. Kappel, F. Micheli, D.R. Brumbaugh, K.E. Holmes, J.M. Mendes, K. Broad, J.N. Sanchirico, K. Buch, S. Box, R.W. Stoffle, and A.B. Gill. 2006. Fishing, trophic cascades, and the process of grazing on coral reefs. Science **311**:98–101.

Navarrete, S.A., E.A. Wieters, B. Broitman, and J.C. Castilla. 2005. Benthic-pelagic coupling and the oceanographic control of species interaction. Proceedings of the National Academy of Sciences USA **102**:18042–18051

Nielsen, K.J., and S.A. Navarrete. 2004. Mesoscale regulation comes from the bottom-up: intertidal interactions between consumers and upwelling. Ecological Letters **7**:31–41.

Paine, R.T. 1966. Food web complexity and species diversity. American Naturalist **100**:65–75.

Paine, R.T. 1980. Food webs: linkage, interaction strength and community infrastructure. Journal of Animal Ecology **49**:667–685.

Pauly, D., Christensen, V., Dalsgaard, J., Froese, R., and F. Torres. 1998. Fishing down marine food webs. Science **279**:860–863.

Pikitch, E.K., C. Santora, E.A. Babcock, A. Bakun, R. Bonfil, D.O. Conover, P. Dayton, P. Doukakis, D. Fluharty, B. Heneman, E.D. Houde, J. Link, P.A. Livingston, M. Mangel, M.K. McAllister, J. Pope, and K.J. Sainsbury. 2004. Ecosystem-based fishery management. Science **305**:346–347.

Pimm, S.L. 1991. The balance of nature? Ecological issues in the conservation of species and communities. The University of Chicago Press, Chicago.

Polis, G.A. 1991. Complex trophic interactions in deserts: an empirical critique of food-web theory. American Naturalist **138**:123–155.

Polis, G.A., and D.R. Strong. 1996. Food web complexity and community dynamics. American Naturalist **147**:813–846.

Polis, G.A., W.B. Anderson, and R.D. Holt. 1997. Towards an integration of landscape and food web ecology: the dynamics of spatially subsidized food webs. Annual Review of Ecology and Systematics **29**:289–316.

Polunin N.V.C., editor. 2008. Aquatic ecosystems: trends and global prospects. Cambridge University Press, Cambridge, U.K.

Power, M.E., D. Tilman, J.A. Estes, B.A. Menge, W.S. Bond, L.S. Mills, G. Daily, J.C. Castilla, J. Lubchenco, and R.T. Paine. 1996. Challenges in the quest for keystones. Bioscience **46**:609–620.

Raffaelli, D. 2000. Trends in research on shallow water food webs. Journal of Experimental Marine Biology and Ecology **250**:223–232.

Redford, K., and P. Feinsinger. 2001. The half-empty forest: sustainable use and the ecology of interactions. Pages 370–399 *in* J. Reynolds, G. Mace, K. H. Redford, and J. G. Robinson, editors, Conservation of exploited species. Cambridge University Press, Cambridge, U.K.

Roberts, C.M., J.A. Bohnsack, F. Gell, J.P. Hawkins, and R. Goodridge. 2001. Effects of marine reserves on adjacent fisheries. Science **294**:1920–1923.

Rodriguez, S.R. 2003. Consumption of drift kelp by intertidal populations of the sea urchin *Tetrapygus niger* on the central Chilean coast: possible consequences at different ecological levels. Marine Ecology Progress Series **251**:141–151.

Ross, D.J., Johnson, C.R., and Hewitt, C.L. 2002. Impact of introduced seastars *Asterias amurensis* on survivorship of juvenile commercial bivalves *Fulvia tenuicostata*, Marine Ecology Progress Series **241**:99–112.

Ruíz, G.M., Fofonoff, P.W., Carlton, J.T., Wonham, M.J., and Hines, A.H. 2000. Invasion of coastal marine communities in North America: apparent patterns, processes, and biases. Annual Review of Ecology and Systematics **31**:481–531.

Seiderer, L.J., and F.T. Robb. 1986. Adaptive features of a bacteriolytic enzyme from the style of the mussel *Choromytilus meridionalis* in response to environmental fluctuations. Actes de Colloques **3**:427–433.

Shafir, A., and J.G. Field. 1980. Importance of a small carnivorous isopod in energy transfer. Marine Ecology Progress Series **3**:203–215.

Steinberg, P.D. 1992. Geographical variation in the interaction between marine herbivores and brown algal secondary metabolites. Pages 51–92 *in* V. Paul, editor, Ecological roles for marine secondary metabolites. Comstock Press, Ithaca, N.Y.

Steinberg, P.D., J.A. Estes, and F.C. Winter. 1995. Evolutionary consequences of food chain length in kelp forests. Proceedings of the National Academy of Sciences, USA. **92**:8145–8148.

Steneck, R.S. 1998. Human influences on coastal ecosystems: does overfishing create trophic cascades? Trends in Ecology and Evolution **13**:429–430.

Thompson, R.C., Crowe T.P., and S.J. Hawkins. 2002. Rocky intertidal communities: past environmental changes, present status and predictions for the next 25 years. Environmental Conservation **29**:168–191.

Troell, M., Robertson-Andersson, D., Anderson, R., Bolton, J.J., Maneveldt, G., Halling, C., and T. Probyn. 2006. Abalone farming in South Africa: an overview with perspectives on kelp resources, abalone feed, potential for on-farm seaweed production and socio-economic importance. Aquaculture **257**:261–281.

Tugwell, S., and G.M. Branch. 1989. Differential polyphenolic distribution among tissues in the kelps *Ecklonia maxima, Laminaria pallida* and *Macrocystis angustifolia* in relation to plant defense theory. Journal of Experimental Marine Biology and Ecology **129**:219–230.

Turner, R.E., and N.N. Rabalais. 1994. Coastal eutrophication near the Mississippi River Delta. Nature **368**:619–621.

Velimirov, B., and C.L. Griffiths. 1979. Wave-induced kelp movement and its importance for community structure. Botanica Marina **22**:169–172.

Vilchis, L.I., M.J. Tegner, J.D. Moore, D. James, C.S. Friedman, K.L. Riser, T.T. Robbins, and P.K. Dayton. 2005: Ocean warming effects on growth, reproduction, and survivorship of southern California abalone. Ecological Applications: **15**:469–480

Vinueza, L.V., G.M. Branch, M.L. Branch, and R.H. Bustamante. 2006. Top-down herbivory and bottom-up El Niño effects on Galápagos rocky-shore communities. Ecological Monographs **76**:111–131.

Walker, B.H. 1992. Biodiversity and ecological redundancy. Conservation Biology **6**:18–23.

Ware, D.M. and R.E. Thompson. 2000. Interannual to multidecadal time-scale climate variations in the northeast Pacific. Journal of Climate **13**:3209–3220.

Wieters, E. 2005. Upwelling-control of positive interactions over mesoscales: a new path linking bottom-up and top-down processes. Marine Ecology Progress Series **301**:43–54.

Wootton, T.J., and M.E. Power. 1993. Productivity, consumers and the structure of a river food chain. Proceedings of the National Academy of Sciences **90**:1384–1387.

Wootton, J.T., M.E. Power, R.T. Paine, and C.A. Pfister. 1996. Effects of productivity, consumers, competitors, and El Niño events on food chain patterns in a rocky intertidal community. Proceedings of the National Academy of Sciences **93**:13855–13858.

Worm, B., and JE. Duffy. 2003. Biodiversity, productivity and stability in real food webs. Trends in Ecology and Evolution **18**:628–632.

Worm, B., H. K. Lotze, H. Hillebrand, and U. Sommer. 2002. Consumer versus resource control of species diversity and ecosystem functioning. Nature **417**:848–851.

Index